Transport and the Environment

45 901 272 7

Transport and the Environment

The Linacre Lectures 1994–5

Edited by

BRYAN CARTLEDGE
Principal of Linacre College
University of Oxford

Oxford New York Tokyo
OXFORD UNIVERSITY PRESS
1996

Oxford University Press, Walton Street, Oxford OX2 6DP
Oxford New York
Athens Auckland Bangkok Bombay
Calcutta Cape Town Dar es Salaam Delhi
Florence Hong Kong Istanbul Karachi
Kuala Lumpur Madras Madrid Melbourne
Mexico City Nairobi Paris Singapore
Taipei Tokyo Toronto
and associated companies in
Berlin Ibadan

Oxford is a trade mark of Oxford University Press

Published in the United States
by Oxford University Press Inc., New York

A catalogue record for this book is available from the British Library

Library of Congress Cataloging in Publication Data

Transport and the environment / edited by Bryan Cartledge.
(The Linacre lectures ; 1994–5)
Includes index.
1. Highway engineering—Great Britain—Environmental aspects.
2. Traffic engineering—Great Britain—Environmental aspects.
3. Transportation engineering—Great Britain—Environmental aspects.
I. Cartledge, Bryan, Sir. II. Series: Linacre lectures ; 1994–5.
TE57.T73 1996 363.73′1—dc20 95-52661 CIP

ISBN 0 19 854934 2

Typeset by Advance Typesetting Limited, Oxfordshire
Printed in Great Britain by Biddles Ltd., Guildford, Surrey

Acknowledgements

It is a pleasure, once again, to record the gratitude of Linacre College, and of the University of Oxford, to British Petroleum plc for their generous sponsorship of the Linacre Lectures. This is the second series of Lectures to be given under BP's auspices; the partnership has continued to be smooth and productive. Although they did not choose or suggest it, the theme of this series is of course particularly relevant to BP's interests and I was glad that a very senior representative of the company, Mr Rodney Chase, was able to deliver one of the Lectures; his visit to Oxford provided the ideal occasion on which to celebrate the BP connection in appropriate style.

I am again grateful to Frances Morphy for putting her editorial expertise at Linacre's disposal and for helping me to shape our contributors' texts into a book; her advice on all aspects of this annual task has been invaluable. Jane Edwards, Linacre's College Secretary, has as usual been indispensable in helping me to arrange the lecture series and in coping with the various last-minute problems that inevitably arise. Finally, I should like to thank the members of Linacre's Governing Body for their support in sustaining this project, which has now become an established feature of Oxford's academic year.

Linacre College, Oxford B.G.C.
August 1995

Acknowledgements

Contents

Authors

Sir Bryan Cartledge, KCMG, MA
Principal, Linacre College, Oxford

Rodney Chase
Managing Director, British Petroleum plc

Michael Dower, MA, DipTP, MRTPI, ARICS
Director-General, The Countryside Commission

Professor Phil Goodwin
Professor of Transport Policy, University College, London

Sir John Houghton, CBE, FRS
Chairman, Royal Commission on Environmental Pollution

Sir Alastair Morton, MA
Co-chairman, Eurotunnel plc

Dr Susan Owens
Lecturer, Department of Geography, University of Cambridge

Sir Bob Reid
Chairman, London Electricity plc and Sears plc; former Chairman, British Railways Board

Dr Hugh Somerville
Head of Environment, British Airways plc

Introduction

Bryan Cartledge

This fifth series of Linacre Lectures was unique both in its topicality and for the sense of urgency that informed the contributions to it.

The issues on which most of these lectures focus are the apparently unstoppable growth of road transport and the environmental impact of this growth: the inexorable rise in the number of exhaust-emitting cars and goods vehicles; the onward march of new roads and motorways to accommodate them; and the distorting dominance of the needs of the motorist in development planning. One of the central conclusions in the report on *Transport and the environment* by the Royal Commission on Environmental Pollution (1994), with which this series of lectures was timed to coincide, is that the present rate of growth of road traffic—it is projected to double by 2025—is environmentally unsustainable. This conclusion has not been seriously disputed. It therefore follows both that the environmental damage caused by the existing level of road traffic must somehow be reduced and that the demand for more road transport must be damped down.

There are no quick or easy means of achieving either objective. As two lecturers pointed out, even the invention of a totally 'green' car would not solve—and might even exacerbate—the problems of traffic congestion and motorway blight. 'Greener' cars, as one element in a package of measures and policies, could nevertheless make a significant contribution to the reduction of atmospheric pollution. Less than 20 per cent of cars and lorries account for 80 per cent of polluting emissions. A determined assault on older and ill-maintained vehicles could thus make a significant impact. Car manufacturers should be given a commercial incentive to design more fuel-efficient engines (a leading Japanese manufacturer has already shown that this can be done): the level of road tax could perhaps be related to fuel consumption per litre of cubic capacity. The efficiency of catalytic converters should also be improved and a date set by which vehicles lacking a converter would be liable to penalty. The traffic-dense countries of Europe should consider following the Japanese example of establishing very stringent—and expensive—road tests for cars that are more than 3 years old; in Britain, the existing MOT test should be much more strictly controlled and much more rigorous.

At the same time, as the Royal Commission and several of our lecturers recommend, steps must be taken to make travel by private car a less attractive option. This is where the real problems begin. Car ownership has dramatically improved the quality of life of millions of people, not least the elderly and the country-dwellers: it has increased their independence, widened enormously the range of choices open to them, and broadened their social and cultural horizons. No government could readily or prudently contemplate negating these benefits. If this country and others are to be spared the nightmare of national gridlock, however, travel by private car must become the exception rather than the rule for the majority. This will require a massive shift in individual attitudes and priorities, comparable to that which within a generation has transformed a nation of, predominantly, smokers into a nation of, increasingly, non-smokers. It could be achieved by a combination of carrots and sticks. The largest *carrot* would be greatly improved public transport, both rail and road. In urban areas, the increased provision of dedicated bus lanes, many more cycle paths, more park-and-ride facilities, and greater frequencies of bus service would help to wean users from the private car; in rural areas, a greatly enhanced network of minibuses could have the same effect. In both town and country, schedules should be planned to provide much greater integration with railway timetables. The most effective *stick* would probably be fiscal: a combination of a significant increase in the duty on fuel—the Royal Commission has proposed an increase of 9 per cent per annum in real terms—with a comparable rise in road tax could be expected to produce some results, although the average car-owner has already shown a remarkable capacity for financial sacrifice for the sake of the freedom which a car bestows.

Car-owners cannot be expected to reduce their dependence on their cars, however, if so much current development planning deliberately reinforces that dependence. The large comprehensive schools and district general hospitals that have been built during the last 3 decades on green-field sites away from town and city centres cannot, obviously, be relocated, but local authorities should be obliged to ensure—as a condition, for example, of licensing private bus companies—that access to them by public transport is greatly improved. The owners of superstores and shopping malls on sites ill-served by public transport should be encouraged to operate, on a commercial basis, their own local minibus services in order to provide a viable alternative to shopping by car; some out-of-town supermarkets already do this as a public relations exercise. Above all, there should be a presumption of refusal on all new planning proposals for developments that presuppose access by private car as the norm. Together, these measures would not only contribute to the slowing of growth in private motor traffic that must be achieved, but would also prevent the social divide (to which some contributors draw attention) between car-owners and the increasingly disadvantaged carless section of the population from widening further. This is a highly desirable objective in itself.

The passenger car is, of course, only a part of the problem. Road freight accounts for a high percentage both of traffic-related air pollution and of trunk road congestion. From every environmental standpoint the railways are greatly preferable as a means of transporting freight, and they have immense spare capacity which could be taken up without any additional investment. Rail freight is not, however, a commercially attractive alternative to the lorry and privatization is unlikely to help to make it so. A massive switch of freight from road to rail is nevertheless essential and, again, both carrots and sticks are called for. The railways must be encouraged to invest in the creation of facilities, at new freight terminals throughout the country, for the rapid transfer of freight from trains to lorries for the last, local leg of its journey. The railways must be able to promise next-day delivery, as the road hauliers already can. The switch to rail could generate savings of up to 80 per cent in energy use and comparable reductions in atmospheric pollution, to which diesel engines are the most dangerous contributors, and there would be a significant bonus in terms of road safety. In Chapter 5 of this volume, Bob Reid argues that, if the railways' share of long-haul freight (over 200 km) were to be doubled, the requirement for road freight would be reduced by 45 per cent. In Chapter 6 Alastair Morton points out that two-thirds of British overseas trade, the proportion that used to be carried in deep-ocean ships, is now borne on the short sea routes to Europe—much of it on lorries. The Channel Tunnel is a new factor in this situation. Initially, it will encourage even more road freight—down the M20 to Folkestone, through the Tunnel on rail transporters, and thence on to the European motorway system and vice versa. But, when the fast rail link to the Tunnel has been completed and the through network to the Midlands and the North of England improved, rail freight to Europe should become much more attractive. According to Dr Brian Mawhinney, former Secretary of State for Transport, the Channel Tunnel could take 400 000 lorries off British roads in 2–3 years' time. The stick to complement these carrots will doubtless have to be, as usual, fiscal: road freight must be made to cost more. This could be done by increasing the road tax on heavy goods vehicles (HGVs)—like other fiscal measures to change the pattern of road use, this would have to be introduced gradually in order to minimize its inflationary effect; and, if motorway charging is to be instituted, this could be weighted heavily against the HGV (HGV traffic would be unlikely to be diverted on to minor roads because of the consequent increase in journey time).

All the foregoing proposals, culled from the chapters that follow, are important and could in combination do a great deal to mitigate the damaging impact of road transport on the environment, but a crucial element is missing—the need for a fundamental shift in Britain's transport policy, particularly so far as the road-building programme is concerned. In Chapter 1 Phil Goodwin demonstrates conclusively that no expansion of our trunk road and motorway system on any feasible scale could contain the swelling demand for road space, partly because—and this is one of his most important contributions to the debate—enhanced supply in itself stimulates greater demand, in the form of induced traffic. There are some indications that this message is at last being heard in Whitehall and Westminster, and that expenditure on new roads and on road-widening schemes such as that for the M25 is being significantly cut back as a result. As Professor Goodwin points out, if that phase of the argument has indeed been won, the next step will be to address the even more difficult question of whether existing road provision should actually be reduced, in order further to discourage the use of private cars. The political difficulties of this approach need no emphasis: public opinion will need time to digest the first stage of the argument before the second is addressed.

The theme of the Linacre Lectures on which this book is based was not 'road transport' but 'transport' and the environment. Most contributors concentrated on the problems posed by the car and the lorry because they pose the most damaging environmental threat. But other forms of transport are by no means environmentally benign. In Chapter 8 Rodney Chase points up the problem of sulfur dioxide emissions from shipping, which is being actively addressed by the International Maritime Organization (IMO), while in Chapter 7 Hugh Somerville gives an encouraging account of the way in which the airline industry is tackling the problem of both noise and atmospheric pollution, particularly emissions of nitrogen oxides (NOx) from jet engines. These problems should be susceptible of technological solutions and international regulation. In complexity and political sensitivity they cannot match the problem of how best to cool the British public's love affair with the motor car.

On 12 June 1995, the Environmental Change Unit and Transport Studies Unit of Oxford University held a 1-day conference on 'Transport and the Environment' that filled Oxford's Sheldonian Theatre. It was addressed, among others, by the then Secretary of State for Transport, the Rt. Hon. Brian Mawhinney, in fulfilment of his undertaking to carry forward the so-called 'great debate' on transport policy initiated by his predecessors. By far the most stimulating, if controversial, contribution of the day came, however, from Professor Carmen Hass-Klau of Germany. There is one dimension of transport policy, she said, in which the British stand head and shoulders above their European partners (anticipatory preening in the audience): *TALK*. Despite all the clever words, 'great debates', and elegant reports, the fact remains (she said) that Britain is between 5 and 10 years behind the rest of Western Europe in every aspect of transport practice. No regular British visitor to Europe can doubt the truth of this. In Chapter 2 John Houghton, summarizing the report of the Royal Commission on Environmental Pollution which he chairs, pays tribute to the cycle-track network in Delft and to the integrated public transport of Zurich. Others single out for praise pedestrianization in Freiburg and provision for cyclists in Gröningen and Basle. The rail networks in France and Germany are more likely than British Rail to attract travellers away from the motorways. It is, of course, true that most other European countries have geographical and historical advantages from which the crowded and topographically complex British Isles do not benefit; it is nevertheless the case that for all too many years successive British Governments have preferred words to action in the field of transport policy. It is unhelpful that the Department of Transport should be regarded in Westminster either as a punishment station or as a staging post for higher things. For Dr Mawhinney it was the latter: he was promoted to be Chairman of the Conservative Party within a month of addressing the Oxford Conference and after only 12 months in the post. We must hope that his successor Sir George Young—who, encouragingly, cycles to work—is still Secretary of State when these lectures appear in print and that he will find time to read them. The issues addressed in this small book are crucial to the future of the British economy and of the British environment: they deserve to be taken very seriously, not merely as fodder for further debate but as a basis for action.

REFERENCE

Royal Commission on Environmental Pollution (1994). *Transport and the environment*, 18th report of the Royal Commission on Environmental Pollution. HMSO, London.

1
Road traffic growth and the dynamics of sustainable transport policies

Phil Goodwin

Professor Phil Goodwin, BSc (Econ), MA, PhD, is one of Britain's leading experts on transport economics and policy. After taking his first degree, in economics, and a PhD in Civil Engineering (Transport Studies) at University College, London, Professor Goodwin joined the Department of Planning and Transportation of the Greater London Council in 1974. He served as Acting Head of the Public Transport Economics Section of that Department and then as Transportation Economics Advisor to the Council. In 1979, Professor Goodwin moved to Oxford, where he joined the University's Transport Studies Unit. He was appointed in 1981 to a University Readership in Transport Studies and to the Directorship of the Unit; in the following year he was elected to an Official Fellowship of Linacre College and, in 1990, to a Professorial Fellowship of the College. In 1996 Professor Goodwin returned to University College, London, to take up the appointment of Professor of Transport Policy there. He is a member of the Standing Advisory Committee on Trunk Road Assessment (SACTRA) and has advised a large number of bodies and institutions, both national and local, on transport matters. Professor Goodwin is the author of numerous articles and reports on transportation and gave evidence to the Royal Commission on Environmental Pollution.

BACKGROUND

One definition of sustainability refers to 'meeting the needs of the present without compromising the ability of future generations to meet their own needs'. The key feature of the definition is its emphasis on a process of change over time, and, in the same spirit, policy can only be understood in terms not of what it *is*, but of where it is going. Thus this chapter is an exercise in interpreting the dynamics of policy evolution.

The context to transport discussions in recent years has been the pervasive role transport plays in everyday life. Stokes (1994) shows that we each spend, on average, just over an hour a day on travel, achieving an overall door-to-door speed of about 20 miles per hour, and travelling 21 miles a day. Forty years ago it was 8 miles a day. Current forecasts imply that our total mileage may be approaching 50 miles a day by 2025 and still be increasing. Partly, this represents a genuine increase in opportunities, but partly it reflects the fact that family, friends, shops, and jobs have moved farther apart, thus reducing the activities we can easily carry out in our own neighbourhoods. It is manifest that most people treasure the enrichment of their lives offered by their own car ownership and are increasingly concerned about the impoverishment of their lives caused by everybody else's car ownership.

The advantages are bought at a cost. Transport activities account for about 30 per cent of all the energy used by final consumers—mostly for road transport, mostly personal rather than freight, mostly by car, mostly not for journeys to work. Transport is the largest sector for which carbon dioxide emissions are expected to increase. It is also a major source of carbon monoxide, nitrogen oxides, particulates, and other noxious emissions. Recent research, reviewed by Reid (1994), shows that transport may be implicated in respiratory complaints, asthma, heart problems, and perhaps cancer. It has other irritating or damaging environmental effects such as noise, vibration, community severance, accidents, stress, and the waste of time, money, and effort due to congestion and unreliable services.

NATIONAL ROAD TRAFFIC FORECASTS AND THE NEW REALISM

In the search for solutions, every 5 years or so the Department of Transport produces forecasts of traffic growth. In 1989 revised national road traffic forecasts were produced (Department of Transport 1989). These had a profound effect on transport policy, because the projected trends, which were consistent with past experience, were inconsistent with what could be supported by the road network. When the forecasts were published, the media reported this as a clever coup by the Department of Transport to win more money from the Treasury for road building. But then there was a quiet period, as local authorities worked out what their own road capacity requirements would be to meet their share of this expected growth. The phrase '83 to 142 per cent traffic growth by 2025' became a mantra of transport policy discussions.

During 1989 and 1990 it became clear that in towns there simply was no way that road capacity could be expanded at rates that would match such growth. The consequence was a matter of arithmetic, not politics. On current trends, vehicles per mile of road could *only* increase, and logically congestion could only get worse (in intensity, or duration, or geographical spread). The 'new realism' (a phrase coined by Goodwin *et al.* (1991), but politically a spontaneous and simultaneous creation of transport professionals working at local authority level) was the recognition that, in towns, supply of road space would not—could not—be increased to match demand; therefore demand would have to be reduced to match supply. This is why, over that period, **demand management** quietly, but quickly, became part of the urban transport policy of every political party.

Urban transport policy *in principle* now nearly everywhere proposes an environmentally friendly package, sometimes justified in economic terms and sometimes by environmental arguments, that usually consists of:

- containment or reduction of the total volume of traffic;
- improved and expanded public transport systems;
- better provision for pedestrians and cyclists;
- pedestrianization, and traffic calming, to reduce the dominance of vehicle traffic; traffic restraint and traffic management, aimed at reducing flows and increased reliability rather than maximizing the throughput of vehicles;
- the control of land-use changes and new development, in such a way as to reduce journey length and car use wherever possible;
- interest in charging people directly for the congestion and environmental damage they cause when using the roads, in order to reduce that damage, reduce traffic, and simultaneously provide the funds to pay for the other parts of the policy.

These policies are explicitly part of European Union and British government objectives, though it must be said that implementation in practice is still lagging far behind agreement on the principles. One of the problems is that these policies are rarely all implemented together, and the omission of some elements reduces or even negates the effects of others: for example, Parkhurst (1995) has shown that, in the absence of good public transport, park-and-ride facilities can provide for an increase in the total volume of traffic, rather than the intended reduction.

THE GROWTH OF TRAFFIC RESTRAINT POLICIES

Other European countries are ahead of us. British transport planners now make a sort of modern pilgrimage to see pedestrianization in Freiburg and

Antwerp and Nuremberg; public transport in Zurich and Grenoble; cycle facilities in Gröningen and Münster; ingenious devices to slow traffic in Velserbroek and Berlin. If we look at the work of Hass-Klau (1990, 1993) and others who have charted the evolution of these policies in Europe over the last 20 years or so, we can see a pattern.

The first stage is some limited pedestrianization of the town centre. This is *always* initially opposed by the retail interests in the centre, who routinely assume that their trade depends on car access, but later this opposition has turned to support in nearly all cases. This is because well-designed pedestrianization, with good complementary public transport services, is good for trade. Therefore, after the pedestrianized area is established, it tends to grow, often as far as some 'natural frontier', such as a medieval town wall, or an inner ring road. Clearly, the larger the restrained area the more necessary it is to make special arrangements for delivery vehicles, residents' cars, and public transport penetration into the restrained area. By now there is so much experience, internationally, on how to make this work that no unsolvable problems are caused in the centre itself.

Outside the central areas, we have seen traffic calming extending throughout residential and suburban areas, and pedestrian zones in local suburban shopping centres. All these developments have proceeded with an increasing recognition by politicians that these are successful, vote-winning policies, but with a continual and worrying pressure of traffic growth and land-use changes outside the controlled area, and sometimes quite serious problems at the margins.

So far, the British experience has followed roughly the same track, but later and hesitantly: improvements are often small and fragmented; the policy context is inconsistent; and in some places the design is cheap and shoddy. If one is too fearful of restricting traffic, the whole idea can be undermined. But, recently, there is at last the beginning of confidence and style, and signs that we may be poised for a very substantial 'catching-up' exercise.

It is not an accident that this new consensus happened first and most swiftly in towns, especially the small- and medium-sized historic cities, where there is the most obvious dissonance between traffic and the urban environment, and a sense of cultural identity and civic pride. But the ideas of traffic moderation that had been born in the towns quickly spilled over into the countryside, where the forecasts of traffic growth are much higher. This had a swift effect. For example, Stokes *et al.* (1992) reported to the Countryside Commission that current trends and the Department of Transport forecasts implied traffic growth rates in the countryside of perhaps 300 or 400 per cent over the next 30 years. As outlined in Chapter 4 of this volume, by Michael Dower, the Countryside Commission concluded that 'demand for road traffic should be managed in such a way that it is consistent with the conservation of the basic countryside resource . . . the countryside can neither afford nor accommodate the predicted growth in road traffic'. This view has not yet been converted into the same degree of consensus about realistic policies as in the towns but, when organizations like the Royal Automobile Club (RAC) seek to explain to their members the case against excessive car use in the countryside, it is clear that something really quite important is going on. There are also some good role models provided by successful progress in some other European rural areas, especially the 'up-market' resorts, though there are also negative examples of attractive country and seaside areas that have, as often in Britain, become overwhelmed by the commercial and traffic pressures of tourism.

The achievement of a new consensus in towns, and the beginnings of one in the countryside, about the unsustainability of projected traffic growth may be described as 'phase 1 of the new realism'. In Britain it is marked by a very wide degree of agreement in principle, a more tardy implementation in practice of some but not all of its elements, and a hope—not yet proved—that our political processes are able to deliver the policies agreed.

THE NEW REALISM, PHASE 2: TRUNK ROADS AND MOTORWAYS

But all this rethinking left the national road network relatively untouched: one could only see controversy, with no signs of consensus or a mechanism that could bring it about. By autumn 1994 it became possible to argue that we were now ready to move to the next stage in the argument, phase 2 of the new realism.

Throughout the early 1990s there had been increasingly intense opposition to road schemes from a quite unusual alliance of those with property to defend and those with nearly no property at all, reinforced by technical opposition from some local authorities. While the motives for people to participate in demonstrations are likely to be more personal than technical, there was nevertheless an underlying technical argument. The opponents of each scheme claim that traffic growth would not be so high if there were better alternatives, and that it is wrong to evaluate one road scheme at a time, in isolation from all the others on the same road, and that expanded trunk road capacity delivers more traffic on to unexpandable local roads than they can cope with. Those claims, albeit often intuitive, all corresponded to an increasingly respectable line of academic research and professional judgement.

In addition, in the summer of 1994 a new, decisive argument emerged from an unexpected source. A seminal report was published by the British Road Federation (Centre for Economics and Business Research 1994). It established, almost by accident, that the imbalance between potential demand and possible supply *is true of trunk roads also*. In the report, McWilliams of the Centre for Economics and Business Research calculated what would happen with the current trunk road programme, of £2 billion a year, and what would happen if the trunk road programme was increased by 50 per cent, to £3 billion a year, from now until the year 2010.

The results, though sensitive to some modelling assumptions that are open to argument, seem robust. They suggested that with the current trunk road programme the congestion on the trunk road network as a whole will get worse every year, not better. And, even if £3 billion pounds a year were spent, congestion would *still* get worse every year. This seemed to confirm something that many were beginning to suspect, but that had *not* been part of the emerging consensus: no realistic trunk road programme can keep pace with forecast traffic growth, on current trends.

That proposition inevitably must have a traumatic effect on transport policy. It may often be reasonable to ask people to accept some personal sacrifice, or environmental loss, if it will make things better overall. But, on current trends and the current programme, things are not going to get better. The sacrifice—at best—just slows down the pace at which things get worse. On trunk roads and motorways, as in towns, it has become necessary to accept that supply of road space is simply not going to expand in line with demand. And therefore on trunk roads also, not just in town centres, demand will have to be moderated to meet supply.

This is the core transport planning axiom of our time, and it will affect *everything* else in the transport sector of the economy. We are now in the foothills of an inexorable rethinking of trunk road policy as far-reaching as that which has largely been completed in the towns. This axiom is also a powerful lever for consensus, because, while there is not yet the basis for agreeing on the scale of road-building, *any* scale is going to need some overall traffic limitation on the trunk roads.

Experience in the towns suggests that, once this is accepted, it is realistic to seek agreement on the methods and objectives of doing so. The axiom entails a logic that will lead to new features in the next stage of discussion about trunk road policy, including *price* (which is the subtext to road-pricing, and to real tolls, though shadow tolls are probably very slightly worse than useless), *engineering and design* (for example, reducing the frequency of access points to motorways to discourage local traffic), and *legal restrictions* (for example, stricter control on the competence of drivers or reliability of vehicles). The balance of these has still to be sorted out. There is also an effect in shifting the function, and therefore style of implementation, of some traffic measures. For example, advanced traffic control and information systems, initially marketed as a form of electronic queue-jumping club for individual subscribers, will instead need to be deployed to keep the actual traffic volume at a level less than the unstable maximum capacity of the road.

The common feature is a recognition that it will be necessary to protect the most important, and most efficient, classes of traffic—primarily freight vehicles and express coaches—by reserved lanes or other priority measures.

Within this policy context, some new capacity may be provided, but according to different criteria. It is not possible to design a new road until it is decided for what traffic load to design it, and that now implies a policy choice, not a forecast. Political authorities now have to choose whether to have their traffic engineers design for 80, or 60, or 40 per cent of potential demand. Any of these choices means that each scheme for extra capacity has to be accompanied by an explicit complementary traffic restraint policy that matches the traffic to that capacity; otherwise, any benefits are rapidly eroded.

For example, the thinking now is that bypasses will be smaller, aimed at diverting through traffic rather than providing for future growth, and with closures or restrictions in the bypassed area to ensure that the planned environmental improvements are actually achieved.

The implication is that we shall see a very much greater emphasis on the quality of the road system, and less on quantity. Expenditure on road maintenance, strength of pavement, design quality, and especially on integrated alternatives to road buildings will *increase*, and expenditure on providing new capacity will *reduce*. There will be some much needed large public transport infrastructure projects, but many more small, local, and widespread improvements to public transport systems, which can give better results than prestige projects.

TRANSFORMATION OF THE VESTED INTERESTS

There is a common view that 'this is all very well, but the vested interests of business, freight operators, the construction industry, motoring organizations, and Department of Transport are too powerful to allow this change in policy'. The question does have to be addressed. The proposition argued here is that the vested interests themselves are becoming reoriented by the same underlying arithmetic of traffic growth.

The vested interests of *business* will need to campaign for reserving priority use for the most important classes of traffic, and providing better alternatives for a decent proportion of the rest. This can easily be common cause for a new alignment composed of freight operators, public transport operators, and environmentalists.

The commercial interests of the *construction industry* will logically shift to maintenance contracts and a substantial programme of redesign and repaving—on trunk roads to provide stronger surfaces for the heavy loads, and in towns to provide materials and design more suitable for streets as living spaces, not just through routes. Quality is not cheap, and there will be very profitable opportunities for those companies that quickly develop the new specialist skills.

The role of the *motoring organizations*, in that scenario, would develop further in the direction of broader 'travellers'' organizations, with more balanced interests in all modes of transport, and indeed in the provision of financial and leisure services that have only marketing connections with car use—an evolution of the 'A' in Automobile Association (AA) and RAC from 'automobile' to 'access', especially now that the original technical interest in the car for its own sake, as a hobby or an icon, is on the wane.

Concerning the *Department of Transport*, it cannot be described as a monolithic vested interest for road building. It is starting to play an important role in encouraging the new policies in towns. The new Highways Agency has responsibilities for managing the road system, not just construction. The Secretary of State for Transport's 'great debate', though noticeably problematic on public transport, has nevertheless provided a breathing space and forum for deeper understanding of the issues of traffic growth and road building than has been the case with any other Minister's pronouncements in recent decades. One particular problem in taking a dynamic view of policy is that it is rarely possible for a government to say 'this is the policy now, but next year it will be different': as soon as that statement is made, the present policy is in effect abandoned. Thus, the process of changing a policy direction is inherently more complicated for a government department to carry out, than for an academic to record—because history cannot be disclaimed, and future policy cannot be pre empted, and there is still some genuine uncertainty and lack of full agreement.

This argument therefore says that the classic role and strength of the big vested interests, and the social attitudes on which they rest, are not immutable, and not *necessarily* a barrier to change. The point is, the writing really is on the wall. The clever business interests and pressure groups realize it. The 'road building alliance', now, is very much softer and more diffuse than it used to be, with all sorts of new strands. Symptomatically, the interests that are not adjusting are starting to adopt the aggrieved tone and marginal political influence of a minority, rather than the confident assurance of the powerful consensus they used to represent.

Such developments are a logically necessary response to traffic growth, and clearly make environmental protection a more achievable objective, since there is scope for a convergence of environmental and economic objectives. There is, though, perhaps some sense of cultural loss. Amateur car maintenance introduced millions of people to the basic principles of engineering and pride in personal mechanical skills in a popular culture that apparently regarded these skills as even more important, for a short period of history, than the basic skills of preparation of food. But it was not a culture that could survive modern car design or Department of Transport (MOT) testing, or a hobby that could survive the disappearance of that nostalgic old phrase, 'the open road'. The 'love affair with the car' is still a powerful image in advertising, but in real life most people have more realistic and modest objectives, to get from A to B without too much fuss.

TOPSY-TURVY ARGUMENTS

Less means more

Policies certainly do not go away just because some fundamental tenet underpinning them has been abandoned, and we are now seeing the perverse creation of some topsy-turvy arguments.

The long-standing assumption was that—as far as possible—the supply of new road capacity should be aimed at providing enough space to cater for forecast demand. It follows that, the higher the traffic forecasts, the higher the required capacity provision.

If it is not possible to match road capacity to forecast levels of traffic, then the relationship between forecast and policy changes. The 1989 traffic forecasts represent a transition from self-fulfilling forecasts to self-defeating ones, or a shift from what Owens (1995) called 'predict and provide' to a new paradigm of 'predict and prevent'. In reaction against this is a line of argument that would say: 'The traffic forecasts were technically faulty anyway: traffic growth has been exaggerated. There will not be as much traffic as we have assumed. Therefore it *will* be possible to provide enough capacity to match it. So we can return to the previous policy.'

With this argument, the lower forecast could justify a bigger road programme than that justified by a higher forecast. Its adherents would hope that the next revised traffic forecasts in 1 or 2 years' time will be lower (which they almost certainly will be, though not for that reason, and not with that implication).

Some aspects of this approach may be detected in a view that is emerging from some sections of the motor industry, notably the International Organization of Motor Vehicle Manufacturers, based in Paris. Streit (1995), for example, proposes that in the mature markets of Western Europe car ownership is already nearly at saturation and little further growth is expected, while the network of highways and motorways should be developed to generate prosperity. In general, forecasts of car ownership and use from the motor industry are lower than those emerging from governments and academics, and there is consequently less inclination to consider environmental objectives as requiring restrictive action.

Induced traffic

The economic and environmental appraisal system for road schemes has been predicated on the assumption that the volume of traffic will be the same with or without road improvements—'road building does not generate extra traffic'.[1]

Faced with declining professional and public confidence in this assumption, the Department of Transport charged its Standing Advisory Committee on Trunk Road Assessment (SACTRA) to investigate and report.

[1] It is unclear how this prevailing orthodoxy was first established. Since the 1930s, the received wisdom was that the quality of the road system *did* influence the volume of traffic. At least one Transport Minister (Leslie Burgin, in 1938) said so, and eminent advisors such as Bressey and Lutyens (1938) and Glanville and Smeed (1958) gave figures to support this view. Foster (1963), shortly before he became a key figure in developing the road appraisal methods of the Ministry of Transport, wrote firmly, 'Build a new expressway and it attracts new traffic on to the roads, which later tends to offset the initial decongestion … (this) must be taken into account in working out the return on road investment'. I have not been able to track down a single piece of empirical evidence published between 1965 and 1975 that could explain the transition from one orthodoxy to its opposite that occurred in that period. It seems as though a procedure of convenience developed in which, at first, such extra traffic was ignored. Over the years this ignoring became an assumption, the assumption became a 'tried-and-tested' practice, the practice became a 'known scientific fact' (with any counter-evidence invariably discounted), and, finally, as a result of a case in the House of Lords in 1980, the scientific fact became a legally enforceable axiom, unchallengeable at road inquiries. I mentioned this puzzle at the Linacre Lecture on 13th October 1994, and acknowledge a number of interesting letters from those present, notably Mr D. Allen of Mouchel & Partners, who suggested that the origin lay with protestors at the early M25 Inquiries who claimed that generated traffic benefits, which the Ministry of Transport had included, biased in favour of the scheme and should be excluded. In retrospect, this carries some irony.

The SACTRA (1994) conclusions, subsequently broadly accepted by the government, were that, on the balance of evidence:

- the construction of extra road capacity does induce additional traffic;
- the induced traffic enjoys some additional benefit itself, but it also slows down the other traffic, shortening the period of relief from congestion offered by a new road; the induced traffic must also almost always increase the environmental damage;
- this erosion of benefit has the worst effect where there is already congestion, because the benefits are substantially reduced due to the non-linear relationship between the volume, and speed, of traffic;
- appraisal of nearly all road schemes should make an estimate of the size and impact of induced traffic.

An appropriate average rule of thumb is that each 10 per cent improvement in traffic speed would cause about 5 per cent more traffic in the short term, and up to 10 per cent more traffic in the longer term, before taking account of the damping effect of that extra traffic on congestion. This corresponds with induced traffic of approximately 10 per cent of base traffic in the short run, for the average road scheme, and 20 per cent after several years, but with a very wide range, from zero to perhaps 40 per cent or so (as, for example, in the case of the M25). Put in another way, when speed increases enable people to save travel time, on average between half and all of the saved time is ploughed back into more travel.

It is understandable that those opposing road schemes have been eager to accept the evidence for induced traffic and that those supporting road schemes have tended to reject that evidence or minimize its scale. But following acceptance of the SACTRA recommendations a new argument emerged, as recorded by the journal *Local Transport Today* (30 March 1995): 'Traffic generated by new roads should be welcomed as a sign of new economic activity, Richard Diment, director of the British Road Federation, told a conference in Taunton last week ... "Many people in many parts of the country would love to see some additional traffic on their roads", Diment said, "as it would be a sign that the economy was picking up".' Foster (1995)[2] pursued a relating line, suggesting 'the right transport

[2] On 23 May 1995 the Institution of Civil Engineers organized a debate for its members, entitled 'The SACTRA report: milestone or millstone'. I proposed 'milestone' and Sir Christopher Foster proposed 'millstone'. I was naturally pleased to win the vote at the end of the debate, by a substantial majority. However, I acknowledge that he expresses a significant discomfort about the apparent tide of opinion against road construction, and fears policies that will make things worse, not better. This discussion is by no means resolved.

answer … to the discovery that there will be too much congestion after a road opens … is to plan a larger road subject to an overall cost benefit test.' Both used the phrase 'there is too little investment, certainly not too much' and argue that therefore we should expand the road programme further—to provide enough capacity to match the forecast traffic levels *and* a bit extra to cater for the induced traffic.

The problem with this argument is that it would imply a new road programme well over 100 per cent bigger than the one that produced such opposition in the 1980s and 1990s, just for the trunk roads, plus even greater increases on all the surrounding local roads. This simply is not going to happen. Therefore, an increasing proportion of the road network will be operating close to capacity, for longer periods of time, and these are precisely the conditions where a small volume of induced traffic has the most damaging effects on traffic congestion. So the caveat 'subject to an overall cost benefit test' is likely to be inconsistent with the proposed strategy: the bigger schemes will be tested and mostly rejected.

UNCHARTED WATERS

All this underlines the point that we are now entering into uncharted waters in transport policy. We simply do not know, with any exactness, what will happen as congestion and pollution intensify or, alternatively, as traffic-moderating policies are implemented. This puts a premium on policies that may be monitored and flexibly adapted in the light of experience, such as pricing and management.

So this is what is meant by 'phase 2 of the new realism'. In the countryside, and on the national network of trunk roads and motorways, we are about to enter into a period of major uncertainty and reappraisal, as we grapple with the implications of the very new recognition of the unsustainability of matching road capacity to the unconstrained forecasts of growth. The outline of the alternative policies is emerging, but is still misty and imprecise.

The pace of change is also uncertain. In the Linacre Lecture from which this chapter was constructed I made the following statement:

> At current pace, this clarification will probably take less than 2 years. If that is right, then the roads programme will not be a very important issue in the next general election, because the transition will have been completed, and there will be as little difference between the parties as there already is on urban traffic. But, if it takes longer, then it could be a more prominent issue, and very divisive. I said a year ago that the M25-widening proposals would not be taken to public inquiry: well,

that is a falsifiable test of the validity of my interpretation of the pace of change in transport policy. I am still of the same view.

Dr Mawhinney, Secretary of State for Transport, announced the abandonment of the M25 proposals in a statement in Parliament on 3 April 1995.

TOO MUCH ROAD SPACE ALREADY?

Now it is time to consider, very cautiously, some questions still in the realm of speculation, but which seem to follow logically once we question how much new road capacity it is sensible to provide. How do we assess the 'correctness' of the *present* road capacity? Is it conceivable that we could already have too much road capacity? Could one imagine a Beeching Report on the road network concluding that some roads do not pay their way? And what would follow from such an assessment?

On the face of it, these are nonsensical questions. 'Everybody knows' that we do not have too much road capacity. Tentatively, five reasons may be offered to question this intuitive certainty.

First, the current level of road provision was constructed on certain assumptions: that there would be *no* induced traffic to erode time savings or cause environmental damage, and that traffic growth would not be constrained, either by mounting congestion or by environmental policy. Suppose those assumptions were wrong. Then some roads that a more correct assessment would have rejected may have been approved. If there are such roads, the current, orthodox appraisal methods would suggest that we would now be better off if they had never been built (providing, of course, that the money had instead been spent on some more rewarding purpose, which is a necessary assumption in rejecting an investment).

The second reason is a curious mathematical anomaly in the science of traffic flow, known as 'Braess's paradox' (Braess 1968). This states that there are certain arrangements of the links in a road network where provision of an extra road results in *increased* overall journey time. Removal of the offending road would make everybody better off. It is not clear that any real-world road has ever been closed for these reasons—County Councils do not employ a team of Braess inspectors looking for anomalous roads—though it is true that road closures planned as part of pedestrianization schemes have often been found to produce traffic chaos that was milder than predicted. So far, the theory is the 'black hole' of transport studies, theoretically important, but difficult to find. It may be time to look again at the real traffic effects of road closures.

The third reason was proposed by Mogridge *et al.* (1985). This conjecture starts from the observation that, in equilibrium, car and public transport must be about equally attractive for the marginal traveller choosing between them. Then a new road increases speeds. This attracts some people to change from public transport to the car, which reduces the speeds somewhat. Then the public transport operator, faced by falling demand, reduces the frequency or increases the fares. This encourages more people to use the car. So far, this describes a vicious circle that has been recognized for decades. However, the nub of the theory is that a new equilibrium will not be reached until the increased traffic congestion has once again made the two modes equally attractive (or unattractive) for the marginal chooser. Since public transport *must* end up in a worse state than the initial conditions, so will traffic speed. Everybody loses.

The main policy conclusion is that to make the conditions of car use better it is necessary to improve public transport: better public transport emerges robustly as one of the most important elements of all the new policy thinking. But what happens if we reverse the logic? Suppose road capacity is *reduced: if* this persuades some people to transfer from car to public transport, and *if* public transport operators then improve services, the new equilibrium will be one in which the attractiveness of public transport is greater and so will be the conditions for car use. Everybody gains if, and it is a strong condition, the dynamic process is reversible.

A fourth reason that we might have too much road space is an economic argument due to Solow (1973). If road users are not charged the full cost of the congestion and environmental damage they cause, there is too much traffic, but also the market for urban land is distorted. As a result, the cost of buying that land, for road building, is less than its full value, which tempts planners to build too many roads in the congested, but underpriced areas. Solow speculated theoretically that this would probably mean that too much road space would be provided near town centres.

That is why the fifth reason is so compelling—indeed, without it all the other arguments would be destined to remain textbook curiosities. By now, there is observation *in practice* of successful and widespread reductions in road capacity in the centre of cities, in connection with their conversion into pedestrian areas. This gives practical confidence that there are some circumstances where policies of reducing road capacity, even in the face of increasing car ownership, are not as illogical as they sound.

Now, what follows from all this? Even if the five arguments are completely valid, it certainly does not necessarily follow that we should rush into a major programme of digging up roads and converting them back into the fields and houses they displaced. Even if the last 20 years were all a mistake—which is not suggested—overturning a mistake rarely, perhaps never, creates the same conditions as not making it in the first place. But addressing this question carefully, before it becomes a big political bandwagon, will help improve our understanding of what we really want from a road system, may help avoid mistakes in the new directions of policy, and may usefully identify *some* situations where reductions in capacity will have a beneficial effect. We should keep an open mind about this possibility.

TRAVEL BEHAVIOUR

Underlying all this debate is the big unsolved question: what will people *do* in response to all these new policies? Transport science has developed elaborate and highly sophisticated methods, and even a significant industry, to explain present choices and predict future ones.

Because transport is so pervasive, and so commonplace, we know that travel choices are related to much broader choices people make about their lives and activities. These other choices constrain the possibilities of changing travel habits in the short term, and give rise to complex and subtle ways of avoiding irksome new policies. We know that it can take several years for the response to a new road, or a change in costs, to work its way through, as old habits are broken and new ones formed. Responses to policies that run counter to well-established trends are unlikely to be faster.

Yet most models and forecasts used in transport studies either claim that all responses are virtually instantaneous, which is not credible, or they attach no explicit time-scale at all to the various future equilibrium outcomes they describe, which is not falsifiable. This static framework seems conceptually wrong, and politically unhelpful, for the assessment of policies whose whole *raison d'être* is to try and break long-established trends.

It can also lead to a systematic bias in the assessment of projects and policies (Dargay and Goodwin 1995). It can no longer be taken for granted that it is necessary or even helpful to rely on the idea of 'equilibrium' and its associated notions of optimizing choices, perfect information, and reversible responses. We were not in equilibrium when the data we use were collected, we are not in such a state now, there is no guarantee that the system is currently moving towards it, we will never arrive there, and, even if we did, we would not stay there for long. For all its importance in the history of economic thought, the concept of equilibrium now acts as a barrier to understanding how things are changing, and how they might change.

Just as transport policy can only be understood by looking at its history, dynamics, and future, so we shall also need new approaches to understand travel behaviour as a process, not a state. We do not yet have an adequate understanding of the process by which travel habits are formed or broken or about the process by which cultural values and patterns of behaviour, in relation to car use, are transmitted from person to person, between producer and consumer, or from generation to generation. We are not even sure about the direction of this transmission, as demonstrated when environmentally conscious children come back from school and pester their parents about environmentally damaging activities.

It is remarkable that in recent years rethinking policy has proceeded faster, and with more successes, than rethinking analytical methods. That is a real challenge for research.

CONCLUSION

In summary, we simply cannot allow traffic growth to continue on its projected path. We will have to accept that the congesting and polluting transport activities will be more expensive and more restrictive. But this in turn provides the opportunity, and the cash, to make the less congesting and less polluting activities cheaper, more convenient, and more attractive than we have ever seen. For that reason, environmental imperatives need not force us into lower standards of living or transport efficiency: rather they can trigger policies that create *higher* levels of welfare and efficiency, but that have in the past been inhibited by market imperfections, unrealistic policy aspirations, and inappropriate understandings and methods of analysis.

ACKNOWLEDGEMENTS

This chapter is a contribution to the research programme of the Transport Studies Unit (TSU) as a designated research centre of the Economic and Social Research Council, whose funding is gratefully acknowledged, as is the endowment from the Chartered Institute of Transport that enabled the TSU to be formed. The chapter has relied on research projects especially supported by the Rees Jeffreys Road Fund, South Yorkshire Passenger Transport Executive, and Department of Transport. I would like to thank BP for sponsoring the lecture and, above all, my colleagues in TSU, Sylvia Boyce, Sally Cairns, Joyce Dargay, Ann Heath, Graham Parkhurst, Gordon Stokes, and Petros Vythoulkas, for their research, ideas, and administrative help.

REFERENCES

Braess, D. (1968). Über ein Paradox der Verkehrsplanung. *Unternehmenstorchung*, **12**, 258–68.

Bressey, C. and Lutyens, E. (1938). *Highway Development Survey 1937*. Ministry of Transport, HMSO, London.

Centre for Economics and Business Research (1994). *Roads and jobs*. British Road Federation, London.

Dargay, J. M. and Goodwin, P. B. (1995). Evaluation of consumer surplus with dynamic demand. *Journal of Transport Economics and Policy*, **xxix** (2), 179–93, May.

Department of Transport (1989). *National road traffic forecasts (Great Britain) 1989*. HMSO, London.

Foster, C. D. (1963). *The transport problem*. Blackie & Son, London.

Foster, C. D. (1995). The dangers of nihilism in roads policy, Henry Spurrier Memorial Lecture, Chartered Institute of Transport, March 1995.

Glanville, W. H. and Smeed, R. J. (1958). Report given on the basic requirements for the roads of Great Britain, ICE Conference on the Highway Needs of Great Britain, 13–15 November 1957. Institution of Civil Engineers, London.

Goodwin, P. B., Hallet, S., Kenny, F., and Stokes, G. (1991). *Transport: the new realism*, report no. 624. Transport Studies Unit, University of Oxford.

Hass-Klau, C. (1990). *The pedestrian and city traffic*. Belhaven, London.

Hass-Klau, C. (1993). Impact of pedestrianisation and traffic calming on retailing: a review of the evidence from Germany and the UK. *Transport Policy*, **1** (1), 21–31.

Mogridge, M. J. M., Holden, D. J., Bird, J., and Terzis, G. C. (1985). The Downs/Thomson paradox and the transportation planning process. *International Journal of Transport Economics*, **14**, 283–311.

Owens, S. (1995). From 'predict and provide' to 'predict and prevent'?: pricing and planning in transport policy. *Transport Policy*, **2** (1), 43–9, January.

Parkhurst, G. (1995). Park-and-ride: could it lead to an increase in car traffic? *Transport Policy*, **2** (1), 15–23, January.

Reid, C. (ed.) (1994). How vehicle pollution affects our health. The Ashden Trust, London.

SACTRA (Standing Advisory Committee on Trunk Road Assessment) (1994). *Trunk roads and the generation of traffic*. HMSO, London.

Solow, R. M. (1973). Congestion cost and the use of land for streets. *Bell Journal of Economics and Management Science*, **4** (2), 602–18.

Stokes, G. (1994). *Travel time budgets and their relevance for forecasting the future amount of travel*. European Transport Forum, PTRC Ltd, Warwick.

Stokes, G., Goodwin, P. B., and Kenny, F. (1992). *Trends in transport and the countryside*. The Countryside Commission, Cheltenham.

Streit, H. (1995). Disaggregate approach to individual mobility demand and the saturation hypothesis. Report delivered to the OICA/ECMT Workshop on long term forecasts of traffic demand, Berlin, May 1995.

2

Sustainable transport: how the Royal Commission sees the future

John Houghton

Sir John Houghton, CBE, FRS, has since 1992 been Chairman of the Royal Commission on Environmental Pollution, whose report, Transport and the environment, *was published while this series of lectures was in progress. He took his first degree, and in 1955 was awarded his DPhil, as a member of Jesus College, Oxford, where he was elected to a Fellowship in 1960. After spending 3 years on the staff of the Royal Aerospace Establishment, Sir John was appointed to a Lectureship, and subsequently to a Readership, in Atmospheric Physics at Oxford University; in 1976 he became Professor in that subject. After four years as Deputy Director and then Director (Appleton) of the Rutherford Appleton Laboratory, he was appointed Director-General of the Meteorological Office and held this post from 1983 until 1991. Sir John is co-chairman of the Scientific Assessment Working Group of the Inter-Governmental Panel for Climate Change and a member of the UK Government Panel for Sustainable Development. He is the author of several books on atmospheric physics, including* Global warming: the complete briefing *(1994), and numerous scientific papers on atmospheric radiation, spectroscopy, and climate change.*

INTRODUCTION

My purpose in this chapter is to present a broad overview of transport and the environment as developed by the Royal Commission on Environmental Pollution during the 2½ years of the preparation of its report, published in 1994. The Royal Commission acknowledges the assistance given to it by many organizations and individual experts including some of the contributors to this volume.

At first the Royal Commission focused on the particular environmental problems posed by transport—for instance, air pollution, noise, and impact on amenity. But we realized very early on that we could not consider those problems in isolation from the more general considerations of transport needs and policy. Our study had to be much broader and to address such questions as why people need to travel, why travel especially by car is growing so rapidly, and what are the links between travel, development, and the life-style of individuals.

Almost everybody is concerned with transport to some degree. The Royal Commission received a mountain of evidence from government departments, from industry, and from a very wide range of organizations and individuals. This evidence contained two main messages. Firstly, motor transport is very important to all of us—it provides for freedom of access and movement, which is basic to our quality of life; many benefits have resulted from the road improvements of recent decades. But, secondly, if the growth of motor transport were to continue as currently envisaged over the next 20 or 30 years, irreversible damage would be done to the environment—such growth is not sustainable. This second message is very widely believed; it was put strongly to us by a wide variety of bodies, including motoring organizations as well as environmental pressure groups.

The evidence setting out the problems posed by transport was voluminous; by contrast very little was offered in the way of solutions to the fundamental dilemma. The challenge presented to the Royal Commission was to propose ways in which the longer-term development of transport can be made environmentally sustainable. That means providing the access people want for continued economic growth and for their livelihoods and leisure, but eliminating—or at least drastically reducing—the many forms of damage that are already too apparent.

The Royal Commission on Environmental Pollution's report was published at the end of October 1994 and immediately provoked a very lively response from the media. On the day of its publication it was the first or second item of news on virtually every news bulletin during the day and it was given high prominence in the newspapers. The issues concern all parts of society and our recommendations clearly touched many nerves.

In this brief account of our report, I will describe the growth of transport, the key environmental concerns, and the extent to which environmental damage may be costed. I will then outline the Royal Commission's approach and the means by which change might be achieved, addressing in particular what is required in towns and cities, in public transport, and in the movement of freight.

THE GROWTH OF TRANSPORT

It has been estimated that the average distance people travel daily has increased by nearly three orders of magnitude over the past 200 years, to

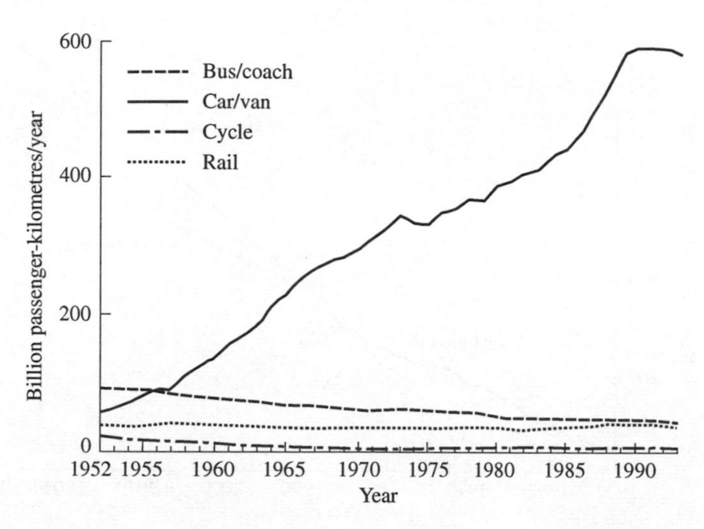

Fig. 2.1 Growth in surface transport in the UK over the period 1952–93: movement of people by various modes of transport. (Source: Department of Transport, 1993.)

about 18 miles a day in the UK. It is interesting to note that the amount of time a typical person spends in travelling tends to remain roughly constant at about 1½ hours per day; it is the average speed of transportation that has grown so much as a result of the enormous developments in the means of transport over this period.

Over the past 40 years, the growth in the surface transport of both people and goods has largely been in motor transport; the movement of people by motor cars has risen by a factor of about ten (Fig. 2.1). Other modes of transport have shown significant decline. Transport overall (defined by the 'gross transport intensity' which relates the amount of movement of goods, people, and carriers) has grown over this period faster than the gross domestic product (GDP). About 20 per cent more 'transport', so defined, is now required per unit of GDP compared with the 1950s. This contrasts with a 40 per cent decrease over the same period in the energy demand per unit of GDP.

The official Department of Transport forecasts of future road traffic (the latest forecasts were produced in 1989) are that it will approximately double by the year 2025 (Fig. 2.2). These forecasts are based on a strong link between road traffic growth and economic growth, the high and the low estimates in Fig. 2.2 representing different levels of economic growth. Traffic growth in the past has tended to follow the high rather than the low forecasts, although since 1989, because of the recession, the actual growth

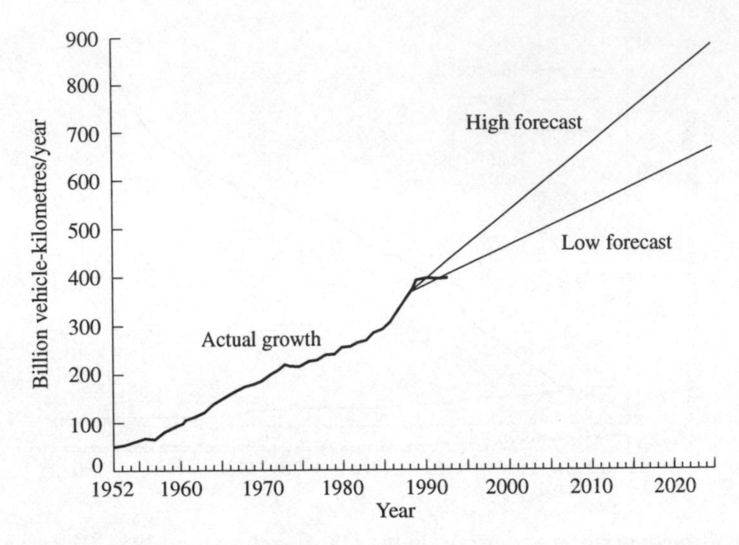

Fig. 2.2 Road traffic growth and 1989 forecasts. (Source: Department of Transport 1989.)

has been closer to the low forecast. It is this level of projected growth that is the cause of the current concern to which we have already referred.

KEY ENVIRONMENTAL CONCERNS

The environmental concerns addressed by the Royal Commission are:

- accidents;
- air pollution;
- the effect on amenity of transport infrastructure;
- carbon dioxide and climate change;
- noise;
- the use of non-renewable resources.

For all of these concerns, road traffic is the principal factor of interest; this is because it is generally the dominant transport mode and also because, for a given amount of transport either of people or of goods, it is generally more polluting than other modes. Transport by air is also of concern; it is faster growing than road transport and in some respects even more a source of pollution for a given amount of transport, but it is only as yet responsible for a fairly small proportion of overall environmental damage. In this chapter I shall concentrate on road transport.

Of the concerns listed above, the ones of which the public are most aware are the first two, both of which directly affect health. The effects of accidents on the people involved are only too apparent. In 1992 road transport accidents accounted for 39 per cent of all accidental deaths in Britain; other transport modes accounted for a further 2 per cent. Despite the great increase in traffic, the number of accidental deaths on the roads has halved over the last 30 years and the trend is still downwards—due largely to efforts by the Department of Transport and to the demanding targets they have set. Nevertheless, the overall numbers remain unacceptably high.

The levels of air pollution in cities in Great Britain often exceed the guidelines laid down by the World Health Organization (WHO). Of particular concern are the levels of nitrogen oxides (NOx) and volatile organic compounds (VOCs) (which are the main precursors to the formation of ozone, the nastiest component of photochemical smog) and of particles small enough in size to penetrate beyond the larynx and enter the lungs.

Air pollution in many cities is highly objectionable to experience, but there is much debate about its lasting effects on health. Although there has been a general increase in the prevalence of respiratory problems among children in recent years, there is no strong evidence that this has been caused by air pollution. However, there is considerable evidence that the symptoms of asthma and other respiratory problems are seriously exacerbated by excessive air pollution. Between 1976 and 1987 acute attacks of asthma increased by a factor of six among children under 5. The Royal Commission on Environmental Pollution (1994, p. 36) has concluded:

> We are concerned that the present use of road vehicles may be causing serious damage to human health by triggering or exacerbating respiratory symptoms and by exposing people to carcinogens from vehicle emissions. The situation should therefore be regarded as unsustainable. Despite the many uncertainties about the effects of transport pollutants on human health and the environment, there is a clear case, on the basis of what is already known, for increasing the precautionary action taken to improve air quality. It is especially important to reduce concentrations of particulates and nitrogen oxides.

Another environmental concern is the loss of amenity due to transport infrastructure, especially the building of roads. Although difficult to specify precisely or to quantify, it was made very clear in the evidence we received that it is a concern that is very much in the public's mind. Developments over the past 40 years in the south-east of England particularly point up the problem. Studies by the Council for the Protection of Rural England (1993) show that areas of the south-east that are not dominated by major roads, airports, built-up areas, or electricity pylons—these they call tranquil areas—have diminished very substantially and the trend of encroachment continues.

Carbon dioxide emitted by transport vehicles, in addition to that emitted from the burning of fossil fuels for the generation of energy and for other industrial purposes, accumulates in the atmosphere and contributes to global warming with the possibility of substantial deleterious impact on climate (Houghton 1994). Concern about climate change due to human activities has risen on the political agenda in the last few years; it is an example of a global pollution problem, in that the carbon dioxide emitted by one country is likely to affect the climate of all countries. At the Earth Summit held at Rio de Janeiro in 1992 the Framework Convention on Climate Change (FCCC) was signed; the first Conference of the Parties was held in March/April 1995 in Berlin. In order to limit the rate of climate change the objective of the FCCC, as stated in Article 2, is to stabilize the concentration of greenhouse gases (that is, those gases that contribute to the greenhouse effect and hence to global warming) in the atmosphere at a level and on a time-scale that 'prevent dangerous anthropogenic interference with the climate system' and that are consistent with the need for sustainable development.

The main greenhouse gas of concern is carbon dioxide which possesses a long lifetime (of the order of 100 years) in the atmosphere. Most developed countries have already set a target that emissions of carbon dioxide in the year 2000 should be no greater than they were in 1990. But, even if this capping of emissions continues after the year 2000, it will not be enough to stabilize atmospheric concentrations; for that, a reduction in global emissions during the next century will be necessary. How big the reduction needs to be and what time-scale is necessary to meet the criteria of the FCCC objective await substantial further research, technical appraisal, and a good deal of political debate. However, because of the long lifetime of carbon dioxide, it is important that action be taken at an early stage. To allow for some growth in the carbon dioxide emissions from developing countries as they industrialize, the target for developed countries is set at reductions of the order of 1 per cent per annum after the year 2000. It is just such a target (in fact, minus 5 or 10 per cent by the year 2010) that was proposed by John Gummer, Secretary of State for the Environment, to the Berlin Conference of Parties in April 1995.

Transport in the UK currently contributes about one-quarter of carbon dioxide (CO_2) emissions; it is also the area of human activity where the emissions of CO_2 are most rapidly growing. In the absence of an increase in fuel efficiency, CO_2 emissions are forecast to grow at the rate indicated by Fig. 2.2. Any overall strategy concerned with the reduction of CO_2 emissions is almost certain to require, therefore, substantial reductions from the expected growth in the transport sector.

Noise from various modes of transport is also a source of concern. For most people road traffic is the most serious cause of noise nuisance although, for those that live near airports, aircraft are a more intensive source of noise. The Royal Commission has made a number of proposals regarding noise reduction; the most effective is to make use of the planning process to ensure that, so far as possible, sources of noise nuisance are kept away from people's homes.

A further environmental concern to do with transport is the use of non-renewable resources in the construction of transport infrastructure and vehicles. A great deal can be done to minimize this through careful design and through the recycling of materials wherever possible.

THE COST OF TRANSPORT

The most obvious elements making up the cost of transport are the cost of vehicles (for example, cars, lorries, and trains) and of transport infrastructure (for example, roads and railway track). These amounted in 1992–93 to an expenditure of about £21 billion on vehicles (of which over £20 billion was on road vehicles) and £7 billion on infrastructure (about £4 billion on roads).

Other costs arise from the effects of transport on people and on the environment; they are costs that transport users impose on others in the community. Many of these costs are very hard to quantify; nevertheless some methodologies for quantification have been developed. For health effects, for instance, the cost of treatment to the health service and of illness to the individual can be estimated. For the effects of climate change, rough estimates have been made of the possible global damage. For noise, information is available regarding what people will pay to avoid, for instance, living near a noisy road. For accidents, recognized although arbitrary allowances for the social costs of deaths and injuries have been used for some years in cost–benefit studies related to safety. Studies carried out in this country and in other European countries have come up with estimates for this group of environmental costs. When aggregated and applied to transport in the UK they mostly fall in the range £11–21 billion per year of which about £6 billion is attributed to the cost

Table 2.1 Tax revenue from road users as a percentage of quantified environmental and public costs

	Costs (£ billion)			Revenue from fuel and excise duty
	Infra-structure	Environ-mental*	Total quantified	(% of total quantified costs)
Cars and light goods vehicles	3.7	7.2–13.3	10.9–17.0	152–98
Heavy goods vehicles	2.8	1.8–3.6	4.6–6.4	68–49

* Including accidents.

of accidents. Most of this (between £10 billion and £18 billion) is attributed to road transport, of which about a quarter relates to heavy goods vehicles.

It is important to emphasize that not all environmental costs can be quantified in money terms, even approximately. Even where the attempt to quantify has been made there are factors that have not properly been taken into account. It can be argued, for instance, that the estimates that have been placed on damage to health or on damage due to accidents do not begin to reflect the human misery or loss that is inevitably associated with such damage. Further, it is virtually impossible to attach a meaningful monetary value to the loss of land for transport infrastructure with the associated loss of access, loss of natural habitats, severance of communities, or visual intrusion that results. These unquantified costs need to be recognized in any attempt at cost–benefit analysis of schemes for transport development.

For road transport a comparison can be made between the total costs to the community, that is, the sum of the cost of providing the infrastructure and the environmental costs, and the tax revenue raised from fuel duty and vehicle excise duty. The figures are given in Table 2.1, from which it will be seen that heavy goods vehicles are very far from meeting the costs they impose. Further, if unquantified environmental costs are taken into account, it is unlikely that cars and light goods vehicles are paying their way. Even more would this be the case if a full economic return on the capital value of the road network were to be demanded of the road user. The conclusion from this analysis is that road transport is not paying for the costs imposed on the community as a whole.

THE ROYAL COMMISSION ON ENVIRONMENTAL POLLUTION'S APPROACH

Given that the result of the economic analysis shows that road users do not pay their way, one approach to addressing the environmental effects would be to propose ways in which the cost of road transport could be increased to the extent required to achieve a better balance. The Royal Commission did not make this the main thrust of its approach for two main reasons. Firstly, as we have pointed out, the economic costings are far from precise and would be difficult to apply in a way that would be seen as fair. Secondly, in order to achieve the appropriate balance it is essential to address particular environmental imperatives more directly.

In its approach the Royal Commission first noted that the basis for a sustainable transport policy was laid down in the government's strategy for sustainable development published in January 1994 (Sustainable Development: the UK Strategy 1994). The purpose of the strategy is to establish a balance among economic development, environmental goals, and the quality of life. The goals for sustainable development in the transport sector are summarized in Box 2.1. The Royal Commission endorses this framework and supports these goals. It is, of course, relatively easy to write them down in general terms. What is much more difficult is to find acceptable ways through which they may be furthered.

The Royal Commission's approach has been firstly to identify a set of eight objectives for reducing the damaging environmental impact of transport and then to propose targets and measures through which the objectives might be achieved. The objectives and related targets are presented in Box 2.2. The Royal Commission believes that the setting of demanding but achievable targets is an essential element of policies that are directed towards reducing pollution and improving the environment.

THE ACHIEVEMENT OF CHANGE

In the space of a short chapter it is not possible to expound all of the Royal Commission's arguments or present more than a few of the 110 recommendations the Royal Commission put forward. I will therefore give some flavour of the recommendations by describing some of the key areas or attitudes where there needs to be substantial change if progress towards more sustainable transport is to be made.

Firstly, our most far-reaching and probably most important recommendations concern decision-making in the transport sector. Large changes

Box 2.1 The government's framework for sustainable development in the transport sector (Sustainable Development: the UK Strategy 1994, p. 69)

The main goal for sustainable development in the transport sector must be to meet the economic and social needs for access to facilities with less need for travel and in ways which do not place unacceptable burdens on the environment. This requires policies which will:

- influence the rate of traffic growth;
- provide a framework for individual choice in transport which enables environmental objectives to be met;
- increase the economic efficiency of transport decisions;
- improve the design of vehicles to minimise pollution and carbon dioxide emissions.

Among the measures available to further these goals are:

- ensuring transport costs reflect the wider costs of transport decisions for the economy and the environment which are not currently priced, and so make transport decisions more efficient;
- land use policies which will enable people and business to take advantage of locations which meet their needs for access with less use of transport or with the use of less polluting means of transport;
- market measures or regulation to improve the environmental performance of transport;
- policies and programmes to promote use of public transport instead of the car, and rail and water to transport freight instead of roads, where these can meet the needs for transport efficiently.

are required in its basis and organization. We want to see better integration of transport decisions with planning decisions concerning development and land use. We also make recommendations that should lead to better co-ordination of transport planning and decision-making at national, regional, and local levels. Even more basic is the need for thorough overall appraisal of all proposals for transport development or improvement. A methodology for such overall appraisal, called the Best Practicable Environmental Option (BPEO), was developed by the Royal Commission on Environmental Pollution (1988). Its basic elements are presented in Box 2.3.

The BPEO procedure or something similar to it might seem rather an obvious way to address policy decisions in a complex field such as transport

Box 2.2 Objectives and targets (Royal Commission on Environmental Pollution 1994, Chapter 14)

A: To ensure that an effective transport policy at all levels of government is integrated with land use policy and gives priority to minimising the need for transport and increasing the proportions of trips made by environmentally less damaging modes.

B: To achieve standards of air quality that will prevent damage to human health and the environment.

B1: To achieve full compliance by 2005 with World Health Organization (WHO) health-based air quality guidelines for transport-related pollutants.

B2: To establish in appropriate areas by 2005 local air quality standards based on the critical levels required to protect sensitive ecosystems.

C: To improve the quality of life, particularly in towns and cities, by reducing the dominance of cars and lorries and providing alternative means of access.

C1: To reduce the proportion of urban journeys undertaken by car from 50% in the London area to 45% by 2000 and 35% by 2020, and from 65% in other urban areas to 60% by 2000 and 50% by 2020.

C2: To increase cycle use to 10% of all urban journeys by 2005, compared to 2.5% now, and seek further increases thereafter on the basis of targets to be set by the government.

C3: To reduce pedestrian deaths from 2.2 per 100,000 population to not more than 1.5 per 100,000 population by 2000, and cyclist deaths from 4.1 per 100 million kilometres cycled to not more than 2 per 100 million kilometres cycled by the same date.

D: To increase the proportions of personal travel and freight transport by environmentally less damaging modes and to make the best use of existing infrastructure.

D1: To increase the proportion of passenger-kilometres carried by public transport from 12% in 1993 to 20% by 2005 and 30% by 2020.

D2: To increase the proportion of tonne-kilometres carried by rail from 6.5% in 1993 to 10% by 2000 and 20% by 2010.

D3: To increase the proportion of tonne-kilometres carried by water from 25% in 1993 to 30% by 2000, and at least maintain that share thereafter.

E: To halt any loss of land to transport infrastructure in areas of conservation, cultural, scenic or amenity value unless the use of the land for that purpose has been shown to be the best practicable environmental option.

F: To reduce carbon dioxide emissions from transport.

F1: To reduce emissions of carbon dioxide from surface transport in 2020 to no more than 80% of the 1990 level.

continued on page 34

Box 2.2 *continued from page 33*

F2: To limit emissions of carbon dioxide from surface transport in 2000 to the 1990 level.

F3: To increase the average fuel efficiency of new cars sold in the UK by 40% between 1990 and 2005, that of new light goods vehicles by 20%, and that of new heavy duty vehicles by 10%.

G: To reduce substantially the demands which transport infrastructure and the vehicle industry place on non-renewable materials.

G1: To increase the proportion by weight of scrapped vehicles which is recycled, or used for energy generation, from 77% at present to 85% by 2002 and 95% by 2015.

G2: To increase the proportion of vehicle tyres recycled, or used for energy generation, from less than a third at present to 90% by 2015.

G3: To double the proportion of recycled material used in road construction and reconstruction by 2005, and double it again by 2015.

H: To reduce noise nuisance from transport.

H1: To reduce daytime exposure to road and rail noise to not more than 65 $dBL_{Aeq,16h}$ at the external walls of housing.

H2: To reduce night-time exposure to road and rail noise to not more than 59 $dBL_{Aeq,8h}$ at the external walls of housing.

where there may be many possible options. The Royal Commission stresses its importance because the procedures that have been employed over the past few decades have not been nearly as comprehensive or complete. Cost–benefit analyses have been carried out for particular schemes, for instance for new roads or for railway investment. These have perhaps been helpful, for example, in enabling comparison to be made between different road schemes of a similar character. But, because of the limited and somewhat arbitrary assumptions on which they have been based, they have been entirely inappropriate for comparison, for instance, between road and rail schemes. Nor have they addressed the wider considerations that pertain to the basic desirability of particular schemes. There is an urgent need for government to develop more comprehensive, thorough, and fair means of appraisal.

Secondly, the report addresses the role of regulation. The most important areas where regulation is appropriate are in achieving the targets for the reduction of tail-pipe emissions and in controlling the use of land for transport infrastructure through the planning process.

Box 2.3 The best practicable environmental option (BPEO) (Royal Commission on Environmental Pollution 1988, p. 14)

BPEO is a systematic and consultative decision-making procedure appropriate to decisions where there are environmental concerns. The BPEO procedure establishes, for a given set of objectives, the option that provides the most benefit or least damage to the environment as a whole, at acceptable cost, in the long term as well as the short term.

Key steps in the procedure are:

- Define the objective in terms which do not prejudice the means by which it is to be achieved.
- Generate options. All feasible ways of achieving the objective should be identified, the aim being to find those which are practicable and also acceptable in terms of their environmental impact and their cost.
- Evaluate the options.
- Summarise and present the evaluation.
- Select the preferred option(s).
- Review the preferred option(s).
- Implement and monitor.

Regarding the control of tail-pipe emissions, more stringent standards are being imposed in stages. The benefits of the catalytic converter will gradually become apparent during the next few years. The Royal Commission is keen to see even more stringent standards after the turn of the century. Of particular importance is the reduction of particulates and nitrogen oxides from diesel exhausts. Some promising technology is becoming available; the promulgation of more stringent standards will ensure its use.

Higher standards are not only needed for new motor vehicles. An urgent matter is the control of emissions from ageing motor cars, lorries, and buses, which produce far more than their share of pollution. Better standards and more effective enforcement need to be introduced to deal with them.

Regarding the connection between transport and planning, there are many recent glaring examples where there has been a failure to consider them together. Both motorways and major developments attract traffic and influence patterns of movement and the viability of other areas (for instance, city centres) far removed from the immediate locations of the new roads or developments. For instance, in the vicinity of motorways and other trunk roads, developments have occurred that in a very short time have led to gross overloading of the road capacity at certain times with the result that the through traffic for which the road was primarily constructed has become subject to heavy congestion. During the last year or two, the Departments of Transport and the Environment have made moves to bring transport concerns and transport provision more centrally into the planning process. These moves are welcome but they require to be strengthened not only by giving stronger guidance but also through the setting up of more appropriate and effective organizational structures for the consideration of planning and development. It is action here that can have the greatest influence on the achievement of sustainable transport in the long term.

Regulation is, however, not the appropriate means for addressing many of the Royal Commission's targets. For these, economic instruments are more appropriate. The Royal Commission considered different measures that are available, including increases in fuel duty, tradeable permits, road pricing, increased parking charges, and support to public transport. The Royal Commission did not give much consideration to tradeable permits (although it did not dismiss their possible use) but it recognized that each of the other measures has a role to assist in meeting one or other of the targets set.

An increase in fuel duty is a simple and effective measure that could assist in achieving several of the targets, in particular a reduction in carbon dioxide emissions from transport, together with the subsidiary target increase in the fuel efficiency of motor vehicles. An increase in fuel duty of about 9 per cent per annum in real terms (compared to 5 per cent which is currently government policy) was proposed. This parallels a recommendation made by the German Council of Environmental Advisors. The technology for increased fuel efficiency is available now: Japanese manufacturers are already marketing average-size motor cars with at least double the fuel efficiency of similar vehicles at present on the road. The elasticity of fuel demand with respect to price is low. We recognize, therefore, that the recommended increase in fuel duty might not in the end be enough, but we considered that announcement of such an increase, especially if other countries did the same, would send a signal to motor manufacturers to develop much more fuel-efficient vehicles and to emphasize fuel efficiency much more in their advertising. It would also provide a strong incentive to car users to choose more fuel-efficient vehicles. We pointed out that, if the increase in fuel efficiency matched the increase in fuel price, the total cost of fuel for the same mileage would remain the same.

There are political as well as economic considerations that affect the general acceptability to the public of such a proposal. Of all the recommendations the Royal Commission made, the fuel price increase was the one that received most criticism from politicians and from the media. Tax increases are highly sensitive politically and it may be that some of the other economic instruments available will have to be employed to achieve the desired ends.

Regarding road pricing (tolls), the Commission recognized its potential power and flexibility as an economic instrument. An important drawback at the moment is that it would tend to divert traffic on to neighbouring roads not subject to pricing. Until road pricing can be attached to all roads of importance in a vicinity, there will be only a very few areas suited to its introduction on a limited scale.

I now move on to particular applications of the approach I have laid out.

Transport in towns and cities

The environmental problems of transport—air pollution, noise, and the disruption caused by traffic to ordinary life—are experienced in their most acute form in our larger towns and cities. Traffic congestion is also generally most severe in towns and cities. To alleviate such congestion it is just not possible, within the limits of acceptable change, to build much more in the way of new above-ground major transport infrastructure. Even if such were to be built it would be likely to create more problems than it would solve.

The targets we have proposed for towns and cities (under objective C in Box 2.2) are focused on less use of motor transport and more use of transport modes that are more environmentally friendly: walking; cycling; the bus; and the light railway. To achieve these targets both negative and positive measures will be required—negative ones (for example, parking restrictions and charges, road pricing) to restrain traffic and positive ones to encourage the other modes.

The advantages of walking and cycling are not only that they are more environmentally friendly but also that, by engaging in them, the general health of people will be improved. But to increase them substantially will require dedicated facilities that are substantially separate from motor traffic. Pedestrianization of areas near city centres has proved to be popular. But, compared with countries like The Netherlands, the amount of dedicated cycle track is very small. During the Royal Commission's visit to Holland, several members cycled around Delft and appreciated the possibility of cycling over tracks for the most part completely separate from motor traffic. A very extensive network is available in that city, where over 40 per cent of the journeys to work (on wet and windy days as well as on fine and relatively still ones) are by cycle and where, for the 5 years before our visit, there had been no fatal accident involving a cyclist. What was clear to us is that a significant increase in cycling in the UK will only occur if dedicated networks are constructed that are sufficiently extensive to be attractive for a variety of journeys. Cycle routes that would enable a much higher proportion of children to cycle to school would be a good beginning.

Many of the Royal Commission's recommendations for towns and cities relate to improvements and developments in public transport, a subject to which I now turn.

Public transport

Many of the Commission's targets and recommendations are aimed at encouraging the growth of public transport as an alternative to private motor transport. Our reasons for this are that, in many cases, public transport is less damaging environmentally and that the use of public transport could be increased substantially without any major extension of transport infrastructure. I cannot here detail the many recommendations regarding public transport developments that we make. What I want to emphasize are three general and important conditions that must be met if a switch to public transport use is to be achieved.

The first is the obvious one that public transport has to be attractive, convenient, reliable, punctual, and easy to use. Different modes of transport, for instance bus and train, need to be well connected, and easily accessible information regarding such connections needs to be provided to potential travellers. In other countries, where there is such a system, it is intensively used. For instance, in the canton of Zurich, which has one of the most efficient surface public transport systems in Europe, public transport usage was 470 trips per person on average in 1990 compared with 290 in London and 130 in Manchester.

The second condition is that public transport schemes have to be viable. Misunderstandings often arise, however, because of a failure to understand the non-linear nature of human behaviour and the responses to transport schemes. One implication of this non-linearity is that inadequate use of one small part of a scheme cannot be employed as an argument that the whole scheme will not be of value. I have already mentioned this in respect to cycling networks. A network will not be used significantly unless it is sufficiently extensive and complete to be attractive and useful. A scheme cannot therefore be properly evaluated until enough of the whole is in place.

A further implication of the non-linear character of transport behaviour is that the combined effect of a number of schemes may be much larger than the sum of their individual effects. As an illustration, I can quote a modelling study carried out for the Department of Transport of the effects

Table 2.2 Effects of different measures on carbon dioxide (CO_2) emissions

Measures	Reduction of CO_2 emissions (%)
New six-line radial light railway system	3.1
Parking charges doubled to £5 per day	3.6
Above 2 measures + £2 central-zone-cordon charge + supply of parking spaces halved	23.3

of various policy measures in reducing the CO_2 emissions from transport (which depend directly on the amount of fuel used). The results when different measures were introduced in a large provincial city are shown in Table 2.2. Although individual measures were quite substantial (for instance, the construction of a six-line light railway), on their own they had little effect in reducing CO_2 emissions. It was only when a number of measures, both positive and negative in character, were put together that a significant reduction in CO_2 emissions emerged.

Further evidence about the effectiveness of combinations of measures comes from cities on the continent of Europe. In Copenhagen, for instance, a city of 1.7 million people, in the early 1970s road-building schemes were abandoned, large numbers of bus priority lanes introduced, and a comprehensive network of cycle paths built. The result has been a 10 per cent fall in traffic since 1970 and an 80 per cent increase in the use of cycles since 1980. About one-third of commuters now use cars, one-third public transport and one-third cycles (*The Independent*, 25 March 1994).

The third condition that is necessary to bring about the large improvement we seek in all aspects of public transport is that there must be real commitment and co-operation on the part of all those involved. One of the reasons why the required degree of commitment has not been forthcoming is that there has been confusion about the respective roles of the central government, local government, and transport operators. It is clearly the role of government to set the framework including much of its detail which is essential to the provision of a user-friendly transport system. It is also the role of government to foster long-term confidence and to ensure that necessary investment is forthcoming—whether from the public or private sector. The Royal Commission recommended that over a period of 10 years there should be a switch of investment from new roads to enable public transport provision to be brought to a standard that meets the conditions of reliability and user friendliness we have mentioned and also to be made sufficiently comprehensive for it to be attractive for a wide range of

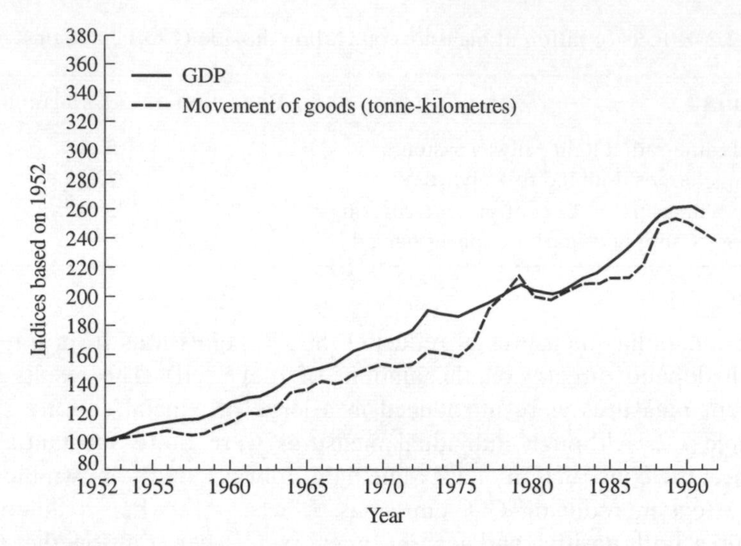

Fig. 2.3 Transport growth in relation to the growth of the economy over the period 1952–93. (Source: Royal Commission on Environmental Pollution 1994.)

journeys. Much careful scientific and technical study will be required to ensure that the best use is made of the available investment.

The transport of freight

Over the last 40 years, freight transport has grown at a rate that closely follows the growth in GDP (Fig. 2.3). Nearly all the increase has been in road transport, while the transport of freight by rail has declined. Much of the increase has occurred because of increases in the average length of freight trips by road.

Freight transport costs are less than 2 per cent of turnover on average—a percentage that has fallen sharply over the last 20 years. Because of this new patterns have emerged. For instance, just in time (JIT) logistics involves precise planning of deliveries to match production or sales needs; by reducing the need for stockholding it has assisted in cutting industrial costs. However, with JIT many more deliveries are made than in a system that depends on higher stockholding. Many deliveries are made by smaller vehicles, which are often not filled to capacity, and in total more fuel is used. Energy consumption can be twice as high in the case of JIT as with conventional logistics.

Freight transport by road is much more environmentally damaging than by rail or by water; in particular, emissions of pollutants are much greater

(Table 2.3). Apart from the pollution they cause, large lorries cause other forms of environmental damage, especially when on minor roads or in towns and cities. Further, as we have already noted, even when low estimates of environmental costs are used, freight transport by road falls far short of meeting its total cost. There are good arguments therefore for making sure that the cost of road freight is increased, which will tend to increase the overall energy efficiency of such transport and encourage the transfer of as much freight as possible to other modes.

How can a modal shift, in particular to rail, be achieved? One of the main attractions of the road for freight transport is its flexibility and convenience; usually the whole journey is possible by road. To encourage transfers to rail, not only has the price to be favourable, but the convenience, flexibility, and reliability of road transport has to be matched as far as possible. This means that adequate investment has to be made in depots and terminals where road/rail shifts can conveniently occur. As there is strong public feeling that the largest lorries should be banned from town and city centres, such terminals should also provide for a shift from large to smaller lorries for local distribution. The reorganization of the railways under privatization could provide big opportunities for imaginative investment in new facilities. A particular opportunity exists for ensuring that there are excellent rail links through the Channel Tunnel with the rest of Europe. There are also opportunities for the more effective use of coastal shipping in the transport of freight over the longer distances.

Many, including government ministers, have recently stressed the importance for the environment of transferring freight from road to other modes. To achieve this will not only require an appropriate level of investment but also the setting up of an integrated policy framework that will ensure that the most effective investments are made.

CHALLENGES FOR THE FUTURE

As I conclude, let me summarize some of the challenges with which we are faced. First of all, some of the environmental problems I have outlined could be eased by improved technology. I have identified four technical challenges that need to be pursued with urgency:

(1) to reduce tail-pipe emissions of pollutants through improvement in catalytic converters and through new technology to remove particulates and nitrogen oxides from diesel exhausts;

Table 2.3 Energy use and emissions of various freight transport modes

Mode of transport	Specific primary energy consumption (kJ/tonne-km)	Specific total emissions (g/tonne-km)			
		Carbon dioxide	VOCs*	Nitrogen oxides	Carbon monoxide
Rail	677	41	0.08	0.2	0.05
Water	423	30	0.1	0.4	0.12
Road	2 890	207	1.1	3.6	2.4
Air	15 839	1206	3.0	5.5	1.4

*VOCs, Volatile organic compounds.

(2) to increase the average fuel efficiency of motor vehicles, the target for motor cars being 40 per cent by 2005;

(3) to develop systems technology to facilitate traffic flow and reduce congestion;

(4) to develop efficient and convenient systems for the intermodal transfer of freight.

Technologies for all of these are basically available. Incentives (through regulation or fiscal means) need to be provided to make sure that they are applied.

In addition to technical challenges, there are substantial organizational challenges that need to be addressed urgently by government:

(1) to develop an effective structure for transport planning at all levels of government (from the central to the local) which is strongly linked to the rest of the planning structure for development;

(2) to introduce appropriate economic instruments to encourage the changes demanded by the environmental imperative. A further increase in fuel duty was recommended in the Commission's report (Royal Commission on Environmental Pollution 1994) as such an instrument. If this proves not to be possible politically, other incentives directed to the same objectives will be required;

(3) to provide a policy and investment framework in which a convenient, user-friendly, accessible, reliable, and 'seamless' system of public transport can be realized by early next century.

To meet these challenges requires strong commitment by everybody involved. By government, who must provide the framework and the incentives. By industry, who must deliver the technology and the systems. By the public, who need to press their demands with a consistency that recognizes both the need for mobility and the desire for a healthy and attractive environment.

Because of the deep and widespread public interest and because of important technical developments the current opportunity for rethinking and planning the long-term future of transport is one that must not be missed. It must be firmly grasped not only in terms of necessary debate but also in terms of deliberate and persistent action.

REFERENCES

Council for the Protection of Rural England and Countryside Commission South-East Regional Office (1993). *South East tranquil areas maps*; CPRE Publications.

Department of Transport (1989). *National road traffic forecasts (Great Britain) 1989.* HMSO, London.

Department of Transport (1993). *Transport statistics of Great Britain 1993.* HMSO, London.

Houghton, J. (1994). *Global warming: the complete briefing.* Lion Publishing.

Royal Commission on Environmental Pollution (1988). *Best practicable environmental option,* 12th report of the Royal Commission on Environmental Pollution. HMSO, London.

Royal Commission on Environmental Pollution (1994). *Transport and the environment,* 18th report of the Royal Commission on Environmental Pollution. HMSO, London.

Secretary of State for the Environment, Secretary of State for Foreign and Commonwealth Affairs, Chancellor of the Exchequer, President of the Board of Trade, Secretary of State for Transport, Secretary of State for Defence, *et al.* (1994). *Sustainable development: the UK strategy.* HMSO, London.

3

'I wouldn't start from here': land use, transport, and sustainability

Susan Owens

Dr Susan Owens was awarded her degree, and subsequently a PhD, in environmental sciences at the University of East Anglia. Since 1981 she has been a Lecturer in Geography at the University of Cambridge and a Fellow of Newnham College. Her long-term research interest is in land use and sustainability; she has written several books and numerous papers on these and other environmental issues. She recently held a Fellowship under the Economic and Social Research Council's Global Environment Change Programme for research on land-use planning and environmental change. Dr Owens has also carried out research for the European Commission, the Department of the Environment, the Organization for Economic Co-operation and Development (OECD), and environmental non-governmental organizations (NGOs). She was a Special Advisor to the Royal Commission on Environmental Pollution during its study of transport and the environment. She is a member of the Technology Foresight Panel on Agriculture, Natural Resources, and Environment and of the UK Round Table on Sustainable Development.

'Relationships between land use and transport are of considerable significance for the environment' (Royal Commission on Environmental Pollution 1994, p. 146).

'... land use planning will be vital if the problems of congestion and energy use are to be solved' (Goodwin *et al.* 1991, p. 114).

'The impact of ever rising levels of transport ... is one of the most significant challenges for sustainable development' (Secretary of State for the Environment *et al.* 1994*a*, p. 169).

INTRODUCTION

As the opening quotations suggest, there is a new awareness of the connections between transport, land use, and sustainability. It is no revelation, of course, that land use and transport interact, nor is concern about the environmental impacts of mobility exclusively a recent phenomenon. But the linking of all three issues, the sense that land-use and transport trends are unsustainable, and a perceptible shift in policy thinking have been apparent for only a few years. This new awareness of interconnected and disturbing trends has been reflected in, and further stimulated by, a series of events and policy developments. Construction of the M3 extension across Twyford Down not only angered and dismayed a wide cross-section of the community but raised consciousness about the land-use and environmental implications of the roads programme; the dawning realization that out-of-town shopping could drain the life from existing centres, as well as adding to traffic and pollution, has opened a vigorous debate on this subject; new Planning Policy Guidance published jointly by the Departments of Environment and Transport (1994) urges planning authorities to use location policy to reduce the need to travel; and the influential Royal Commission on Environmental Pollution (1994, p. 158), in a long-awaited report on transport and the environment, devotes a whole chapter to land use and states unequivocally that: 'Land use planning is an important component of policies designed to reduce the environmental effects of transport.'

This chapter explores connections between land use, transport, and sustainability, considers their social, environmental, and political implications, and suggests some policy prescriptions. The framework is provided by three dichotomies that are recurring themes in the transport debate: accessibility and mobility; needs and demands; consumers and citizens.

CONNECTIONS

Land use and transport connect across a range of temporal and spatial scales: this is precisely why they are so interesting in the context of sustainability. Decisions in particular localities can affect local, regional, and ultimately global environments in the immediate future and in the longer term.

The direct effects of transport systems on land are perhaps the most obvious connection. Roads occupy about 3.3 per cent of the land area of Britain (as much as a fifth in urban areas) (Royal Commission on Environmental Pollution 1994). More significantly, transport infrastructure has major visible impacts on both urban and rural landscapes and serious effects on habitats. These have been among the most important factors in raising the public profile and political prominence of transport issues. In 1992 English Nature considered that the roads programme threatened

150 Sites of Special Scientific Interest and even the Department of Transport conceded that 48 sites were at risk (Royal Commission on Environmental Pollution 1994). It is one of the ironies of current appraisal systems that land of low economic value has a magnetic attraction for roads planners when by its very nature it may be of considerable environmental significance. More invidiously, motor vehicles have expropriated social space. Towns and cities, and increasingly rural areas, are full of them, moving and stationary. Though discrete areas may be reclaimed for part of the time, vehicles dominate elsewhere, and their domination tends to be exclusive of social interaction.

We experience these direct effects daily. The indirect effects of transport on land use may be less apparent but their environmental implications are also severe. Of the 90 million tonnes of primary and secondary aggregates used annually in Britain, one-third is destined for the maintenance or construction of roads. A not insignificant proportion of freight (some 5 per cent) can be accounted for by materials for road building—an interesting example of transport consuming itself (Peake and Hope 1993).

Perhaps of even greater significance for sustainability are the more subtle and longer-term interactions between transport and land use. Transport systems clearly influence patterns of development (there is ample historical evidence for this) and land-use patterns in turn affect travel behaviour. These effects are neither simple nor deterministic—land-use and travel patterns must be interpreted within their social and economic contexts—but it helps clarify the connections to consider how these patterns have evolved over several decades.

Growing prosperity has been associated with increasing mobility of both people and goods, permitting geographical dispersal of residential areas and other land uses. At the same time, the centralization of many facilities has been predicated upon the assumption of mobility—fewer and larger schools, shops, and hospitals have been located where they cannot practicably be reached by means other than car or lorry. In combination, these trends have led inexorably to the increasing separation of homes, jobs, and services. New roads are demanded to serve new patterns of activity and, in turn, transport infrastructure generates its own development pressures, leading to changes in land use and further traffic generation. The effects of this upward spiral of mobility and dispersal are all too evident, in dormitory suburbs and villages, in out-of-town shopping and leisure centres, and in splendidly isolated business parks, where the car is the overwhelmingly dominant mode of travel to work.

Of course, there are social and cultural factors involved in these changes and it would be simplistic to suggest that particular patterns of land use determine people's travel behaviour. But there is a growing body of empirical evidence to suggest that where things are and how they are related to transport networks—especially those land uses that generate a great deal of travel—do make a difference to travel patterns. In low-density suburbs, people travel further and more frequently than their counterparts in more compact urban areas: National Travel Survey data reveal that travel demand rises quickly as densities fall below 15 persons per hectare and falls sharply as they increase above 50 persons per hectare (ECOTEC 1993). Peripheral and out-of-town shopping facilities are associated with quite different travel patterns from those found at central or edge-of-centre sites. One study found that 95 per cent of customers reach a freestanding supermarket in outer London by car, while only a third had travelled by car to an inner London store and 50 per cent had arrived on foot (Shaw 1992). Another suggests that 93 per cent of employees in out-of-town office sites travel to work by car (cited in Royal Commission on Environmental Pollution 1994). There are many more examples. What is most worrying about the spiral of increasing mobility and dispersal is that it is making car use a matter of necessity rather than choice: patterns of high mobility have become, almost literally, set in concrete (Smith 1994).

In terms of the first dichotomy—accessibility and mobility—it is clear that land use and transport have interacted in such a way as to make motorized mobility the predominant means of gaining accessibility. Broadly speaking, travel is a means to an end (increasing congestion perhaps even makes the 'leisure drive' less attractive): people want to live in attractive residential areas whilst at the same time having access to a range of jobs, services, green spaces, and other opportunities for recreation. Such choices are denied to those without access to a car as the spiral of mobility and dispersal has made land-use patterns difficult to serve by any other means of transport. Even when journeys are short (and most still are), the dominance of traffic and its expropriation of otherwise attractive spaces makes walking or cycling an unpleasant, if not daunting, prospect.

The trends that increasingly concern us are usually attributed to a range of social demands—for living and working space, for travel, and for goods and services that can be provided more cheaply when advantage is taken of green-field sites and economies of scale. In terms of the second dichotomy, these demands have been equated with needs, not only in markets that make no distinction between the two but in public policies such as those on provision of new roads. We might expect markets, and in some cases policies, to respond to demands that we have expressed as consumers. But we are surprised and dismayed by congestion, landscape degradation, declining air quality, and the increasing tyranny of traffic. The third dichotomy is relevant here: as consumers we equate demands with needs and set about satisfying them in apparently rational ways; as citizens, we recognize different sets of social needs and obligations (Sagoff 1988) and perceive that our ability to meet them is threatened by current trends in land use and transport.

In outline, these are the ways in which land, transport, and environment are connected. Recognizing these links, and their evolution over time, suggests a number of important implications.

IMPLICATIONS

The most obvious implication—now widely agreed—is that current trends are unsustainable. It follows that we need a clearer idea of what *is* sustainable and that, in trying to attain it, land-use and transport planning cannot be treated as separate processes. A third implication is that the definition and planning of a sustainable transport future must involve confrontation with difficult political and philosophical issues.

The spiral of mobility and dispersal is environmentally unsustainable because of the range and severity of impacts, now well-documented (for example, Banister and Button 1992; Commission of the European Communities 1992; Goodwin *et al.* 1991; Royal Commission on Environmental Pollution 1994; Transnet 1990), and because technical fixes will not be enough to reduce them to acceptable levels. Improvements in energy-efficiency and emissions characteristics of vehicles, for example, are in part absorbed by an addiction to speed and performance but to a large degree simply outstripped by the growth of traffic (Secretary of State for the Environment *et al.* 1994*b*). Even using the Department of Transport's low forecasts of traffic growth, substantial increases in carbon dioxide emissions from transport are predicted and emissions of many other pollutants, though falling initially under the influence of tighter emissions standards, are likely to resume an upward trend towards the end of the first decade of the next century (Royal Commission on Environmental Pollution 1994). If technological improvements cannot cope adequately with pollution from motor vehicles, they have even less to offer in terms of other major problems such as landscape degradation and the dominance of motor vehicles in urban and rural environments. The 'clean green car' is not a solution to the environmental problems of transport.

Trends are also unsustainable economically because—environmental impacts aside—congestion imposes heavy costs on society and we cannot afford to build our way out of it. That attempts to do so are likely to be self-defeating has been acknowledged by the Standing Advisory Committee on Trunk Road Assessment (SACTRA). In an important report SACTRA (1994) confirms what critics of the roads programme have long asserted—that new roads not only redistribute but generate traffic, sometimes very substantially.

Land-use and transport trends are unsustainable socially because they divide communities and inhibit social interaction in urban spaces. More alarmingly, they are leading to the polarization of society into those with and without access to cars. The latter group—which is a substantial one (some 30 per cent of households) in spite of the frequent assumption that it is somehow residual—suffers not so much from lack of access to a car *per se*, but from deterioration in local services that can no longer compete with car-oriented facilities. This is a serious new dimension of social deprivation.

Finally—and this accounts more than anything else for the emergence of a 'new realism' (Goodwin *et al.* 1991 and Goodwin, this volume)—current trends are becoming unsustainable politically. Transport is a sector of the UK economy in which 'almost everything has gone wrong' (Pearce *et al.* 1993, p. 150) and, though prescriptions differ widely, virtually no one is satisfied with the current system. Indications of crisis include growing concerns about the links between traffic pollution and health, complaints about noise, and dissatisfaction with the extent to which traffic dominates townscapes and people's daily lives (for example, Birmingham City Council 1993; Vidal 1994). Traffic projections and the roads programme have become a focus for this discontent, reflected in the existence of more than 200 protest groups opposing new road schemes in the UK (Ghazi 1994). Significantly, the opposition spans a broad political spectrum.

The consensus that it would be intolerable for current trends to continue points to the need to gain a clearer understanding of what patterns of land use and mobility might be sustainable in the longer term. On almost all definitions, sustainability implies living within our environmental means. Defining what is sustainable, therefore, must entail thinking about the kind of environment that we want to enjoy and to maintain for our descendants: it is really a question, as Ulrich Beck (1992, p. 28) puts it, of 'how do we wish to live?' This reasoning leads us in the direction of sustainability constraints and targets and anticipates one of the prescriptions considered at the end of this chapter.

The co-evolution of land use and transport implies that planning should treat them as an integrated whole. Though land-use and transport planning have paid lip service to each other, they have in many important respects been treated as separate processes. The almost irresistible development pressures stimulated in some areas by new trunk roads, for example, have been taken into account neither in estimating the benefits of the roads nor in the development plans on which they are superimposed. Similarly, the transport and environmental implications of locational decisions in retailing, health, education, and other services have been neglected. If transport systems are to serve patterns of development, rather than the other way around, there is a need for more genuine integration of land-use and transport policies at all levels: this too anticipates an important prescription.

If trends are unsustainable, it will be necessary to re-define the relationship between accessibility and mobility—the first dichotomy. We need to find ways of achieving ends with means other than those of ever-increasing movement. The analogy with energy is interesting: it took a long time to realize that economic growth and growth in energy consumption were not necessarily linked by a deterministic 'iron law'. The energy intensity of economic activity in the UK (the energy consumed for each unit of gross national product (GNP)) has decreased over several decades. We still, however, think in terms of iron laws when it comes to transport and economic growth, and certainly these do not yet seem to have been decoupled (Peake and Hope 1993). Such decoupling will be essential for a sustainable transport future and land-use planning has a potentially important role in enabling us to achieve more with less.

The second dichotomy—needs and demands—takes us into more contentious territory. The question here is whether trends in land use and transport are a powerful expression of 'what people want' and by implication would be difficult (even undemocratic) to modify or reverse. Breheny (1995, p. 91), amongst others, has argued that much if not all of urban decentralization can be explained by householders and businesses 'voting with their feet'. Similar views on mobility and dispersal are frequently expressed by decision makers, usually associated with a strong implication that to change current trends would require '. . . draconian policies, with very serious social and economic ramifications' (Breheny 1995, p. 92).

The 'deep-seated trends' argument sees patterns of land use and mobility as the inevitable outcome of the expressed preferences of people and firms for freedom of movement and decentralized locations, and assumes that the resulting development patterns meet such aspirations. Both premises are questionable on several grounds. First, on close examination, many 'deep-rooted trends' can be seen to have been stimulated by actions of the state which in turn may reflect powerful corporate interests. Investment criteria favouring roads, tax relief on company cars, and permissive planning policies are all quite deliberate choices that have helped to drive the spiral of increasing mobility and dispersal. Second, the choices that are manifest in current trends have been made in a context in which many of the costs, including environmental costs, are externalized. Thus market failure, as well as the influence of particular policies, contributes to the blunting of real costs, and influences the evolution of land-use and travel patterns.

It might be argued that, even if the spiral is in part policy-driven, such policies themselves nevertheless reflect 'consumer choice' (Breheny 1995). Even if this is true, however, it is not axiomatic that development patterns that reflect *consumer* choice ultimately meet the aspirations of individuals or represent what the community would consciously choose if the consequences were anticipated. Location and transport choices have the characteristics of many-person 'prisoners' dilemmas' (Parfit 1984), where rational self-interest does not produce the best outcome, even for each individual, and planning is needed in the interests of efficiency. More importantly, perhaps, and returning to the third dichotomy, there are circumstances in which our preferences as citizens for social and environmental outcomes can be satisfied only by policies that counter our immediate preferences as consumers (Owens 1995*a*; Sagoff 1988). Indeed, if consumer choice were sovereign, we would not have land-use or transport planning policies at all.

PRESCRIPTIONS

It is widely agreed that in future such policies will have to change. Before considering particular prescriptions, a brief history of policy developments at the interface of land use and transport helps to set current choices in perspective.

There is nothing new in the recognition that land use and transport interact: indeed, in the 1960s it became fashionable to employ large-scale land-use/transportation models in planning, though these proved rather cumbersome, expensive, and inflexible. Nor have the impacts of transport on the environment only recently been discovered: from the earliest days of motorized mobility, environmental effects in one form or another have been an issue (Smith 1994). Connections between land-use change and the environment have similarly been recognized over a long period. What is new is a much more holistic conception of development, movement, and environment as a complex and interconnected system demanding integrated policy treatment.

The first glimmers of this recognition probably came during the 1970s, when a succession of energy crises, including an abrupt fourfold increase in the price of oil in 1973, focused attention on the energy dependence of patterns of mobility and of urban and rural development. The environmental issue of the day, soon after publication of *The limits to growth* (Meadows *et al.* 1972), was resource depletion rather than pollution or sustainability of the biosphere: concern was about 'running out' of fossil fuels, reflected in the literature in a spate of articles about 'squeezing spread cities' and 'imploding metropolis' (for example, Downs 1974; Franklin 1974). Even so, exhortations to adopt more energy-efficient patterns of development never really crossed the threshold of political legitimacy. In the mid-1970s, when the Select Committee on Science and Technology recommended that the energy implications of all planning policies be explicitly identified (House of Commons Select Committee on Science and Technology 1975), the Government's response was that these were not sufficiently important to be separately considered (Department of Energy 1976). In 1976, a transport minister, responding to a question about the role of land-use planning in limiting the demand for travel, replied that this was 'hardly a question that you can expect me to answer' (quoted in Smith 1994, p. 91).

Nearly 20 years later, a significant shift in thinking is reflected in Planning policy guidance: transport (PPG 13). Not only is PPG 13 an inter-departmental publication—a significant development in itself—but it recognizes both the need for demand management and the contribution of land use planning in this context (Department of the Environment and Department of Transport 1994, para. 1.3):

> By planning land use and transport together in ways which enable people to carry out their everyday activities with less need to travel, local planning authorities can reduce reliance on the private car and make a significant contribution to ... environmental goals ...

Specific recommendations are made about planning and location policies to help achieve these ends (Table 3.1). PPG 13 also has something to say—albeit in a carefully qualified manner—about non-pollution impacts of transport. New routes are to be kept away from Areas of Outstanding Natural Beauty and Sites of Special Scientific Interest 'wherever possible'; in National Parks and Special Areas of Conservation (designated under the Habitats Directive (Council of the European Communities 1992)) they will be allowed only in 'exceptional' and 'strictly defined' circumstances, respectively.

PPG 13 is both a product of and a contribution to the 'new realism' in transport policy (Owens 1995*b*). Its formal recognition of links between land use, transport, and the environment might be attributed to a number of factors, including global warming (which has been used to the full to promote otherwise desirable policies) and, more cynically, the appeal of

Table 3.1 Reducing the need to travel: policies recommended in PPG 13

Development plans should aim to reduce the need to travel, especially by car, by:

- influencing the location of different types of development relative to transport provision (and vice versa);
- fostering forms of development that encourage walking, cycling, and public transport use.

To meet these aims, local authorities should adopt planning and land-use policies to:

- promote development within urban areas, at locations highly accessible by means other than the private car;
- locate major generators of travel demand in existing centres that are highly accessible by means other than the private car;
- strengthen existing local centres . . . that offer a range of everyday community, shopping, and employment opportunities, and aim to protect and enhance their viability and vitality;
- maintain and improve choice for people to walk, cycle, or catch public transport rather than drive between homes and facilities that they need to visit regularly;
- limit parking provision for developments and other on- or off-street parking provision to discourage reliance on the car for work and other journeys where there are effective alternatives.

land-use planning as a long-term and relatively painless instrument as opposed, for example, to significant price increases or serious traffic restraint. A sustainable land-use and transport policy will require much more, however. Most importantly, it will be necessary to establish sustainability constraints and targets, to integrate land-use and transport planning at all levels, to define principles of location policy to reduce the need for environmentally damaging travel, and to apply these principles consistently as part of a cohesive policy package.

Sustainability constraints

Identifying environmental capacities and constraints is a crucial part of the task of defining what is sustainable (Jacobs 1994; Owens 1994). The concepts are already familiar: we should not exceed critical loads for pollution, put pressure on renewable resources beyond their maximum sustainable yield, or cause irreversible harm to important landscapes and habitats. The definition of such constraints will always be a political (and a learning) process rather than an exact science, and their translation into law and practice will be contested. But we cannot identify sustainable land-use and transport policies without some idea of our 'environmental means' and of the constraints and targets that must be defined if we are to live within them. Only a few examples can be mentioned here to illustrate the relevance of these issues in the context of transport and land-use planning.

Emissions reduction targets are not unfamiliar in international and national environmental policy, but there is now a need to disaggregate them to the transport sector and in some cases to disaggregate them spatially to geographical areas that are meaningful in the planning context. Policies proposed for Air Quality Management Areas, intended 'to secure for the first time the effective co-ordination of all activities which can influence air quality improvement . . . in those areas where it is most needed' (Department of the Environment 1995, p. 6), are a step in the right direction. They should provide an opportunity not only for spatial disaggregation of targets, but also for land-use planners to co-ordinate their policies with those more traditionally concerned with air pollution control.

The need to protect important habitats and landscapes from transport developments is also a familiar aspect of policy rhetoric, though in PPG 13, as elsewhere, it is always qualified by reference to 'circumstances' in which such protection might be overridden. The problem is that the definition of such circumstances has tended to be elastic and the 'balancing' of environmental and other considerations on a case by case basis leads inexorably to erosion of what is most precious in the natural and cultural environments. This will continue unless we are prepared to make 'critical environmental capital' effectively inviolable (Owens 1994). And, lest stringent protection of certain environmental assets be interpreted as a licence to despoil the rest, sustainability might also entail a 'no net loss' of landscape or habitat rule. These conditions simply formalize a view that society's 'need' for environmental quality is at least as valid as its 'need' for transport.

Less conventional targets that might help focus land-use planning policies and decisions could include those for proportions of people and freight using different types of transport by particular dates. The Royal Commission on Environmental Pollution (1994) has proposed, for example, that the proportion of passenger-kilometres carried by public transport should increase from 12 per cent in 1993 to 20 per cent by 2005 and 30 per cent by 2020. Such targets might be criticized on the grounds that they are unrealistic or that there are no specific mechanisms for their delivery. But targets in other areas (such as road safety and health) have provided an invaluable focus and a set of goals, and, unless targets are challenging, they represent little more than rhetoric with business as usual.

Integration of land-use and transport planning

Integration of land-use and transport planning is a prescription acclaimed by almost everyone but whose implications in practice often remain obscure. It must mean, however, at least three things. First, that the land-use implications of transport and transport infrastructure proposals are always considered. It is not acceptable for development pressures associated with new roads simply to be superimposed on existing development plans: they should be considered as part of the planning process. Second, the transport and consequent environmental implications of development plans and planning decisions should automatically be taken into account. PPG 13 should help to establish this as a standard part of the planning process. Neither of these steps is easy. Relationships between land use, transport, and the environment are neither simple nor deterministic and it is certainly not the case that planners can plug policies or development proposals into convenient models that will provide the answer in terms of longer-term effects on land-use or travel patterns. But planners are accustomed to living with uncertainty. What is important is that the interactions should be considered as much and as far as possible: undesirable consequences are then less likely to be left out of account altogether. The third requirement relates to mechanism. Undoubtedly, the integration of land-use and transport planning must involve some institutional change. Whether this entails merging departments or closer co-operation and functional integration is perhaps not the most important question. What really matters is breaking through bounded rationalities in both land-use and transport planning: changing the thinking, for example, that has implicitly assumed car ownership in locational decisions or defined transport problems, and therefore solutions, in predominantly engineering terms.

Principles

A considerable amount of research, both theoretical and empirical, now points to a set of principles for land-use planning that could reduce the need to travel and encourage the use of environmentally friendly modes. It should be stressed that these are indeed principles that have to be interpreted in specific localities rather than blueprints for ideal development patterns. The principles concern the location and the separation of activities and the nature of the environment in which movement takes place.

A key principle is to locate developments that generate a great deal of movement in places where they are accessible by foot, bicycle, and public transport. Self-evident, perhaps, but we have only to look at the inaccessible locations of so many businesses and services to see the extent to which this principle has cheerfully been ignored. It is at least now recognized in PPG 13 (see Table 3.1), drawing in part on a well-established tenet of Dutch location policy (Netherlands Ministry of Housing, Spatial Planning and Environment 1991). Indeed the Dutch Environment Ministry may itself be testimony to the effectiveness of sensitive land-use planning policies: when it relocated its previously more dispersed headquarters to a site adjacent to the central station in The Hague, the proportion of employees travelling to work by car fell from 43 to 4 per cent (M. Post, Netherlands Ministry of Housing, Spatial Planning and Environment, 1994, personal communication).

A second key principle is reducing the physical separation of activities, in other words, reversing the spiral of mobility and dispersal. Much has been written about this. Broadly speaking, it implies mixed, accessible development within existing urban areas rather than dispersed, widely separated activities in peripheral or rural locations: it does not necessarily mean compact cities, high densities, or moratoria on development elsewhere.

A third principle is that of 'selective accessibility'. Making developments accessible on the map is not the same as making them accessible on the ground. Even short journeys by foot or bicycle may be frustrating and unpleasant, if not downright dangerous. A vital element of integrated land-use and transport planning at the 'micro' scale is to provide environments in which journeys by the most environmentally friendly modes are consistently the most convenient and attractive: it is these modes that should have the direct routes, priority at junctions, shortest waiting times, and, for bicycles, plentiful, secure, and convenient parking facilities. Such policies are precisely of the kind that pose a challenge to bounded rationalities.

Planning as part of a policy package

Much of the rhetoric about land use, transport, and the environment confuses the need to travel with the inclination to do so. Adopting certain location and development policies (assuming that this is politically feasible) can influence the relative location of different land uses: it may indeed be possible over time to reduce the physical separation of homes and jobs, or homes and services, amenities, or whatever. But this is not the same as reducing the amount of travel by car. Whilst the latter remains the most convenient, prestigious, and, at least at the margin, the cheapest way to travel, people will use it quite rationally to bypass their carefully located local facilities and to expand their (immediate) choice. (Such trade-offs predate motorized mobility: marriage distances increased substantially when bicycles became widely available (Perry 1969).)

Planning can change morphology but not motivation. To modify the latter requires reinforcement by other policy measures. It means that the costs of travel by car must rise, requiring the use of fiscal measures. And it requires investment in alternative forms of transport. In particular, arrangements for the provision of public transport cannot be left to the vagaries of the market (such as it is, in transport): if planners are being urged to locate development close to public transport nodes, they need some degree of certainty that the latter will still exist when their plans come to fruition. Having recognized that land-use planning is indispensable in the evolution of a sustainable transport system, we must not load the planning system with transport and environmental objectives that it cannot possibly deliver unaided by a wider policy package.

CONCLUDING COMMENTS

Once again, these prescriptions may be linked to the dichotomies identified at the beginning of this chapter. In a future where lower *mobility* must be achieved if we are to live within our environmental means, integrated land-use and transport planning can be seen as an important means of maintaining *accessibility* and (longer-term) choice. When land-use planning is set within a framework of sustainability constraints, it can be seen as positive and permissive, rather than negative and restrictive.

Planning must become part of a policy package that reduces the *demand* for transport as well as the *need*. More fundamentally, however, the identification of sustainability constraints challenges the way in which needs and demands have been conceptualized in relation to land use, transport, and the environment. Demand for transport has conventionally been regarded as synonymous with need, affording it a kind of objective status. Demand for environmental quality (which seems to grow with GDP at least as much as transport does) has in contrast been regarded as subjective, if not frivolous. Perhaps we should re-examine all of our demands—for clean air, for foodstuffs from distant places, for rapid journeys, for a rich natural and cultural environment, and so on—and consider whether they constitute 'needs' and whether all are comparable on an equal basis. Such reflection is likely to expose the conflict between consumer and citizen in each of us.

In its policy document on air quality, the Department of the Environment (1995, p. 17) offers a conventional definition of sustainability in the context of land use and transport: 'Sustainable Development means providing for development, and for transport whilst at the same time taking account of the need to protect the natural and built environment'.

Thus demands for development and transport must be 'provided for', while it is sufficient to 'take account of' environmental needs. A reappraisal of needs and demands for development, transport, and the environment might challenge this implicit hierarchy and suggest a re-definition: 'Sustainability means meeting the imperatives of environmental protection whilst at the same time taking account of the need for access'.

It will not be easy for land-use planning to contribute to a sustainable transport future after several decades in which trends have been moving in an unsustainable direction. Ideally, of course, we 'wouldn't start from here', but as the Royal Commission on Environmental Pollution (1994) points out, we have no choice but to start with the physical infrastructure that already exists. The challenge for sustainable land-use and transport planning is to permit our values as citizens to take precedence over our immediate preferences as consumers. The priority now must be to find ways of articulating and expressing these values in the planning process.

ACKNOWLEDGEMENT

The author is grateful to the Economic and Social Research Council for supporting a 1-year Global Environmental Change Programme Fellowship (1993–94) permitting her to explore issues of land-use planning and environmental sustainability.

REFERENCES

Banister, D. and Button, K. (1992). *Transport, the environment and sustainable development.* E. & F. N. Spon, London.

Beck, U. (1992). *Risk society: towards a new modernity*. Sage, London.

Birmingham City Council (1993). South Birmingham Study Public Opinion Survey (Summary). Birmingham City Council, City Engineers Department, and Department of Planning and Architecture, Birmingham.

Breheny, M. (1995). The compact city and transport energy consumption. *Transactions of the Institute of British Geographers, NS*, **20** (1), 81–101.

Commission of the European Communities (1992). *The future development of the common transport policy: a global approach to the construction of a community framework for sustainable mobility*. Commission of the European Communities, Brussels.

Council of the European Communities (1992). *Directive on the conservation of habitats and of wild fauna and flora* (Directive 92/43/EEC). Council of the European Communities, Brussels.

Department of Energy (1976). *Energy conservation*. HMSO, London.

Department of the Environment (1995). *Air quality*. HMSO, London.

Department of the Environment and Department of Transport (1994). *Planning policy guidance: transport*. HMSO, London.

Department of Transport (1989). *National road traffic forecasts (Great Britain) 1989*. HMSO, London.

Downs, A. (1974) Squeezing spread city. *New York Times Magazine*, **38**, 17 March.

ECOTEC (1993). *Reducing transport emissions through planning*. HMSO, London.

Franklin, H. M. (1974). Will the new consciousness of energy and environment create an imploding metropolis? *American Institute of Architects Journal*, August, 28–36.

Ghazi, P. (1994). Brake now, Minister, or there will be tears. *The Observer*, 20 February.

Goodwin, P., Hallett, S., Kenny, F., and Stokes, G. (1991). *Transport: The new realism*. Report to the Rees Jeffreys Road Fund. University of Oxford Transport Studies Unit, Oxford.

House of Commons Select Committee on Science and Technology (1975). *First report on energy conservation*. HMSO, London.

Jacobs, M. (1994). *Sense and sustainability*. Council for the Protection of Rural England, London.

Meadows, D. H., Meadows, D. L., Randers, J., and Behrenv III, W. W. (1972). *The limits to growth*. University Books, New York.

Netherlands Ministry of Housing, Spatial Planning and Environment (1991). *The right business in the right place*. Ministry of Housing, Spatial Planning and Environment, The Hague.

Owens, S. (1994). Land, limits and sustainability: a conceptual framework and some dilemmas for the planning system. *Transactions of the Institute of British Geographers, NS*, **19** (4), 439–56.

Owens, S. (1995*a*). The compact city and transport energy consumption: a response to Michael Breheny. *Transactions of the Institute of British Geographers, NS*, **20** (3), 381–4.

Owens, S. (1995*b*). Predict and provide or predict and prevent?: pricing and planning in transport policy. *Transport Policy*, **2** (1), 43–9.

Parfit, D. (1984). *Reasons and Persons*. Oxford University Press, Oxford.

Peake, S. and Hope, C. (1993). *Transport policy analysis: an energy analogy*, Management Studies Research Paper No. 14/91. Department of Engineering (Judge Institute of Management Studies), University of Cambridge.

Pearce, D. W., Markandya, A., and Barbier, E. (1993). *Blueprint 3: measuring sustainable development*. Earthscan, London.

Perry, P. J. (1969). Working class isolation and mobility in rural Dorset 1837–1936: a study of marriage distances. *Transactions of the Institute of British Geographers*, **46**, 121–41.

Royal Commission on Environmental Pollution (1994). *Transport and the environment*, 18th report of the Royal Commission on Environmental Pollution. HMSO, London.

SACTRA (Standing Advisory Committee on Trunk Road Assessment) (1994). *Trunk roads and the generation of traffic*. HMSO, London.

Sagoff, M. (1988). *The economy of the earth*. Cambridge University Press, Cambridge.

Secretary of State for the Environment, Secretary of State for Foreign and Commonwealth Affairs, Chancellor of the Exchequer, President of the Board of Trade, Secretary of State for Transport, Secretary of State for Defence, *et al.* (1994*a*). *Sustainable development: the UK strategy*. HMSO, London.

Secretary of State for the Environment, Secretary of State for Foreign and Commonwealth Affairs, Chancellor of the Exchequer, President of the Board of Trade, Secretary of State for Transport, Secretary of State for Defence, *et al.* (1994*b*). *Climate change: the UK programme*. HMSO, London.

Shaw, D. (JMP Consultants Ltd) (1992). Traffic impact study for a retail store. Paper submitted to the fourth annual TRICS Conference, September 1992.

Smith, J. (1994). *The politics of environmental conflict: the case of transport in Britain 1972–1992*. Unpublished PhD thesis, Department of Geography, University of Cambridge.

Transnet (1990). *Energy, transport and the environment*. Transnet, London.

Vidal, J. (1994). No visions please, we're British. *The Guardian*, 4 March, supplement p. 14.

4

Roads and road transport: their impact on the countryside—and what should be done about it

Michael Dower

Michael Dower has been Director-General of the Countryside Commission since 1992. He took his MA at St John's College, Cambridge, and then studied for a Diploma in Town Planning, which he was awarded in 1957 at University College, London. He then held a succession of posts in planning and development: with the London County Council, the Civic Trust, the United Nations Development Programme for Ireland, and the Dartington Amenity Research Trust and Dartington Institute in Devon, of which he was Director for 17 years, until 1985. Mr Dower then served as National Park Officer to the Peak Park Joint Planning Board until he was appointed to his present post. He has been a member of the Sports Council and of the English Tourist Board, and he was President of the European Council for Village and Small Town from 1986 to 1990. He is the author of several books and articles about the British countryside and its preservation.

INTRODUCTION

I am glad that this series of Linacre Lectures includes contributions by Phil Goodwin, formerly of the Transport Studies Unit in Oxford, who with his colleagues has contributed to the Countryside Commission's thinking on the subject, and by Sir John Houghton, Chairman of the Royal Commission on Environmental Pollution, whose report, *Transport and the environment*, published in October 1994, has made a powerful contribution to the rapidly evolving public and political debate on the subject of transport. I have shaped my contribution to be complementary to theirs.

The countryside

I should first state my standpoint. The Countryside Commission is a government agency, with a statutory brief to protect and enhance the natural beauty of the English countryside, and to help people to enjoy it.

Thus, I speak for the countryside—that splendour of England, nearly 90 per cent of its land surface, green, full of nature and beauty, but also long settled and much used by people. The countryside carries a special resonance for the people of England. This helps to explain why people come out from the towns in their millions to enjoy the countryside, why many city people aspire to move their homes out into the rural areas (indeed two million have done so over the last 20 years), why over two million people belong to the National Trust, which this month celebrated its centenary, why the protection of the countryside arouses so much passion in so many.

What qualities in the countryside underlie these impulses? Our surveys suggest that people value the peace and quiet of the countryside, the contact with wildlife, the release from the noise and bustle of the city, and the space available to pursue activities of widely varying kinds. People do not want to see the quality of the countryside diminished. This is an 'old idealism' to be set alongside the 'new realism' about traffic that Phil Goodwin describes in Chapter 1.

Therefore, when I consider the impact of roads and road traffic on the countryside, I am talking not about some pristine area of wilderness, but rather a complex zone of already varied and continuous human use. The impact falls upon the place, with its qualities, and upon the people who live and work in it or who visit it for their enjoyment. Moreover, those people themselves generate a significant part of the road traffic that affects the countryside, though the larger proportion comes from interurban travel.

A backward glance

The Countryside Commission is active on transport and traffic because we, with our joint remit to conserve the countryside and to promote public enjoyment of it, are both directly in the line of fire and part of the artillery. Most visitors to the countryside come by car, contributing to the impact on the place and on other users, such as those who like to come by bicycle or on foot. It is in fact one of the ironies of road transport that the campaign for the creation of modern roads was effectively started by a group whose needs are now so inadequately met by those roads, namely, the cyclists. In the nineteenth century, as the railways spread, so the horse-drawn coaches disappeared, the turnpike trusts went bankrupt, road maintenance virtually ceased, and the cyclists found the going impossible. Through their two rival organizations, the National Cyclists Union and the Cyclists Touring Club, they formed in 1886 the Roads Improvement Association, whose first aim was to improve the techniques of repairing roads (Jeffrey 1949).

Over the following 20 years, parliamentary and public pressure grew, together with the numbers of motor vehicles, until Lloyd George in his Budget statement of April 1909 submitted proposals for raising money to be placed at the disposal of a new central authority for making: '. . . grants to local authorities for the purpose of carrying out well planned schemes for widening roads; for straightening them; for making deviations round villages; for allaying the dust nuisance . . . (and for) constructing absolutely new roads' (quoted in Jeffrey 1949, p. 23).

After a difficult passage through Parliament, the Road Board was set up; and from that starting point came a modern road system—tarmacadamed, standardized, classified, equipped, widened, and extended to the point that the motorist could drive without stirring dust to the door of virtually every home or building in the land.

The creation of that road system made it possible for people to realize the dream of the contented motorist, expressed by Toad of Toad Hall in Kenneth Grahame's *The Wind in the Willows:* 'The poetry of motion! The real way to travel! O bliss! O poop-poop! O my!' Yet this reverie was provoked, as early as 1908, by the shattering impact of a motor car upon the horse and cart in which Toad had been travelling. In subsequent years, the number of motor vehicles grew faster than the space on roads, and so the convenience, the safety, and pleasure in travel all became gradually eroded.

The official responses to this have been many and varied. Roads have been widened and improved; new roads have been built; different sorts of traffic have been, to some degree, segregated; some separate cycleways have been built; road surfaces, road crossings, and road vehicles have been made safer; ribbon development has been largely stopped; conscious effort has gone into the design of street furniture, the landscaping of motorways, and even the maintenance regimes of roadside verges in the countryside.

But we have not solved the problem. The congestion is getting worse. There is less and less direct pleasure to be got from motoring. Yet people use cars for more and for longer journeys, because they are seen as a relatively safe and convenient means of travel, while millions of bicycles rust in garden sheds because it is dangerous and unpleasant to ride them.

THE IMPACT OF ROADS AND ROAD TRANSPORT ON THE COUNTRYSIDE

Until the last 2 years, the official view was still that a major part of the answer to this dilemma of congestion was to build or widen yet more roads in order to accommodate not only the existing traffic but also the continuing growth in traffic that the Government predicted. In Chapter 1 of this volume, Phil Goodwin refers to the national road traffic forecasts of 1989 which predicted an overall doubling of traffic between 1988 and 2025.

The Countryside Commission has become increasingly concerned about the impact of roads and road traffic upon the countryside. The prospect of a doubling of this traffic filled us with alarm. The Government's report (Department of Transport 1989) did not indicate how much of the traffic growth implied by the forecasts might fall upon the roads in the countryside.

Therefore, in 1991 we commissioned Oxford University's Transport Studies Unit (TSU) to prepare an analysis, *Trends in transport and the countryside*, which was published in mid-1992 and marked a turning point in thinking about the impact of traffic on the countryside.

By re-analysing the Department of Transport's (1989) official forecasts of road traffic growth, the report showed how the predicted increases in road traffic severely threatened the fabric and the quality of the countryside. It revealed the massive rise in traffic that was likely to occur in rural areas—far more massive than that in urban England—if the official forecasts were in fact realized. The 1989 forecasts were based on an average of urban and rural traffic combined. With these unravelled, it could be shown that *rural* traffic could grow by three or four times in the next 30 years, compared with only 50 per cent growth in urban areas. The simple premise was that, if the forecast of a doubling of all traffic were to come true, then *either* urban densities would have to decline substantially on the American model *or* traffic would spill out to invade the wider countryside on a massive scale.

In publishing the TSU report and (alongside it) our own policy conclusions (Countryside Commission 1992), we made it plain that we do not regard road traffic as an evil. We acknowledge the boon that good road transport represents for people who live and work in the countryside, for those who visit it for recreation or tourism, and for those who pass through rural areas on their way from town to town. But we made it starkly clear that the time had come to acknowledge also the *costs* caused by road transport of the types and scale that now prevail, and by the infrastructure created to serve this transport.

These costs are already heavy, and growing. I would express them by reference to four adjectives:

- *human* costs, in the misery, tension, and waste of energy through accidents, crowding, congestion, and noise, and in the personal isolation and inconvenience caused to those who do not own or have use of cars by forms of development that rely on the use of cars;
- *financial* costs, as in the estimated £18 billion value of the national road-building programme, the costs of maintaining road networks, and the cost of accidents, put by the Department of Transport in 1991 at £750 000 for each road fatality and £25 000 for each serious injury;
- *physical* costs, in the impact of transport systems, and particularly of roads and road vehicles, upon the landscape and cultural quality of many areas and upon the quality of life in many communities;
- wider *environmental* costs, including insidious amenity damage in open country from signing, kerbing, lighting, and so on, habitat destruction, and noxious vehicle emissions.

To illustrate something of these physical and environmental impacts, I offer (Fig. 4.1) two maps of south-east England, one from 1960 and one from 1992, showing the extent of the built-up areas and of the zones or corridors of noise or other significant disturbance caused principally by roads, as well as by major airports, railways, power stations, and high-voltage power lines. The areas of the countryside that are not so affected are defined as 'tranquil areas' by the Countryside Commission and the Council for the Protection of Rural England (CPRE/Countryside Commission 1993), who jointly commissioned the preparation of these maps by the ASH consultancy. The tranquillity to which we refer is precisely the quality that people seek in the countryside, a sense of space and quietude, a release from the noise and bustle of the city.

The contrast between the two maps shows how the tranquillity of the south-east countryside has been shattered over the last 30 years. This effect has been caused not mainly by new built development (which, though substantial, has been quite sharply contained by planning policy, and particularly by green belts), but rather by construction or improvement of major roads, and particularly the four-fold increase in road traffic over that 30-year period. Toad's eulogy of 1908 is becoming the nightmare of the 1990s. No one suggests that we go back to the horse and cart. But roads and road traffic are having a devastating effect already on large parts of our countryside, and we need to find a new approach.

A NEW APPROACH

The sheer scale of the costs that I describe led us to conclude (Countryside Commission 1992) that the countryside simply cannot accommodate nor afford continued traffic growth on the scale implied by the Department of Transport's forecasts of 1989.

The ruinous change brought by roads and road transport to our countryside and the invisible poisoning of the air we breathe through vehicle emissions (not to mention their effect on climate change, which will further affect the countryside) create a pressing national policy dilemma. They imply that we cannot base transport policy simply upon the needs of the road user. We must take account also of the wider public and the wider environment, and strive for a new perception of road transport.

In the 1992 position statement, the Countryside Commission expressed the view that a new approach was needed to tackle the national transport problem. This must be based on two clear policy assumptions.

- There must be an assertive change from making supply meet the constantly rising demand for road traffic, to managing that demand within acceptable limits—the 'new realism' to which Phil Goodwin refers in Chapter 1.
- The engine of this change in transport policy must be the idea of sustainability, as carried through from the Bruntland report to *This common inheritance* (Secretary of State for the Environment *et al.* 1990), and the UN Rio de Janeiro Conference. For us in the Countryside Commission, sustainability means that the quality of the countryside should not be diminished.

We were therefore much encouraged when the government, in *UK strategy for sustainable development: consultation paper* (Secretary of State for the Environment *et al.* 1993), emphasized the major adverse impact of road transport upon the environment and stated the aim of producing a sustainable transport policy. In this paper John Gummer, Secretary of State for the Environment, launched a public debate on this and on

> . . . the range of measures which could contribute to reduced impacts from the sector, including fiscal and market measures to increase the price of transport and reflect wider costs; regulations, such as additional vehicle standards or better speed enforcement; land use changes to improve accessibility and reduce the need to travel; traffic management and public transport improvements.

We in the Countryside Commission were among the many who contributed to that debate, which led to the production of a chapter on transport

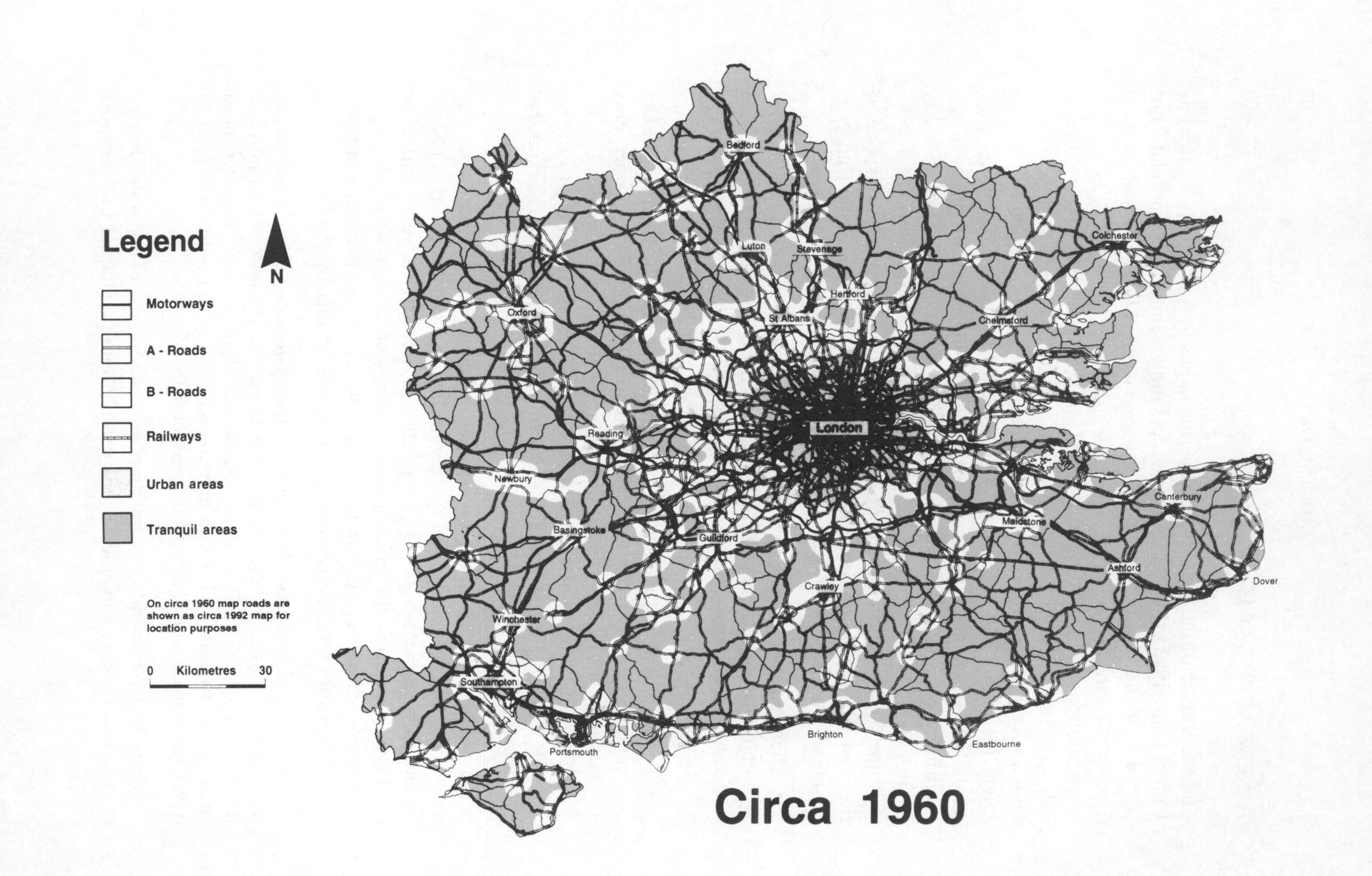

Legend
N
Motorways
A - Roads
B - Roads
Railways
Urban areas
Tranquil areas
On circa 1960 map roads are shown as circa 1992 map for location purposes
0
Kilometres
30
Bedford
Luton
Stevenage
Colchester
Hertford
Oxford
St Albans
Chelmsford
London
Reading
Newbury
Canterbury
Basingstoke
Guildford
Maidstone
Ashford
Dover
Crawley
Winchester
Southampton
Brighton
Eastbourne
Portsmouth

Circa 1960

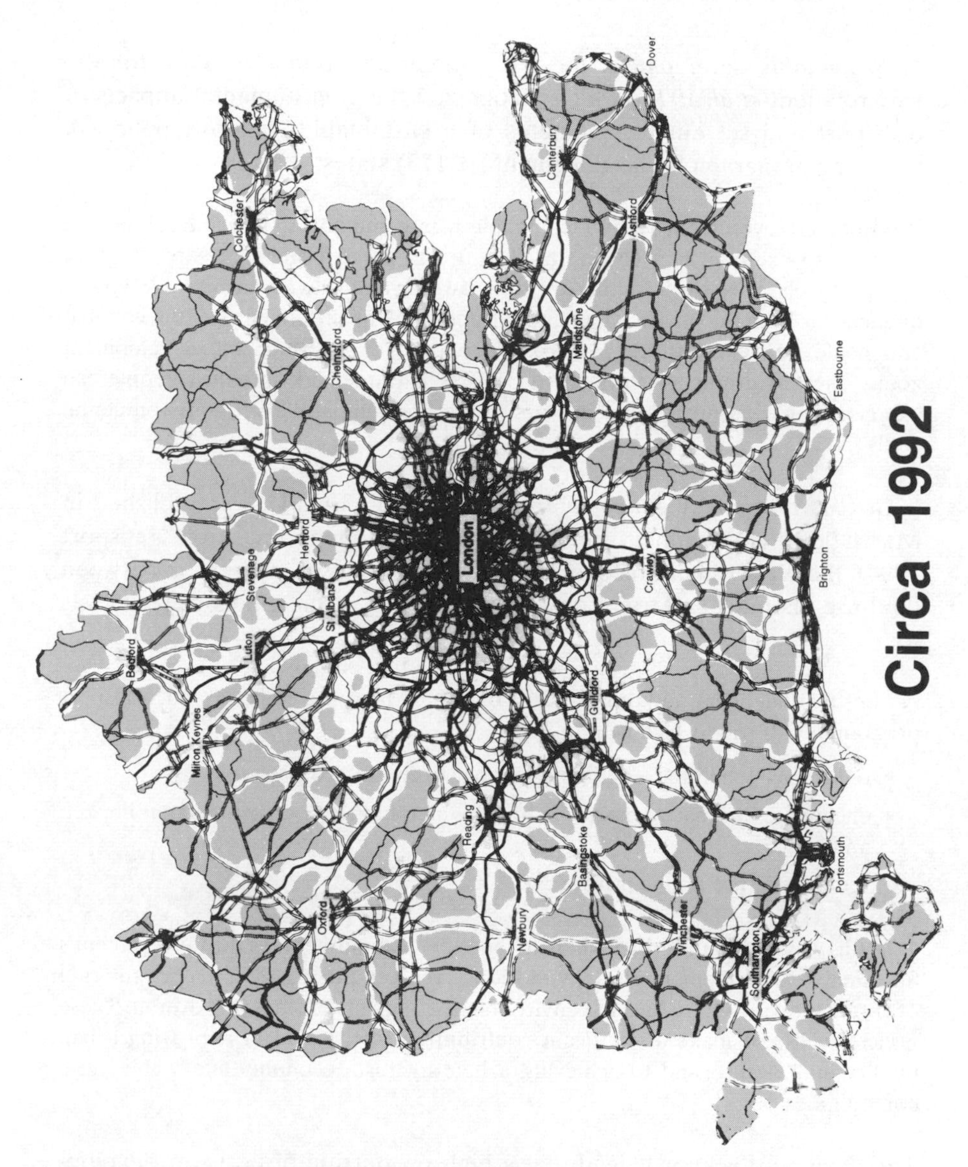

Fig. 4.1 Maps produced by CPRE demonstrate the effect on the countryside of the seemingly inexorable growth in traffic, roads, and new development over the last 30 years. The maps show how south-east England's 'tranquil areas' (areas sufficiently far away from road traffic and development to expect to be able to experience peace and quiet in the countryside) have been shattered. They indicate the need for changes to planning and transport policy that better respect environmental priorities. At current rates of loss, tranquillity will virtually be a thing of the past in the south-east in a matter of decades. (Source: CPRE/Countryside Commission 1993.)

in *Sustainable development: the UK strategy* (Secretary of State for the Environment *et al.* 1994), which recognized the environmental impacts of transport and set out the elements of a sustainable transport policy. A revealing paragraph in this document (p. 173) states:

It is not the Government's job to tell people where and how to travel. But if people continue to exercise their choices as they are at present and there are no other significant changes, the resulting traffic growth would have unacceptable consequences for both the environment and the economy of certain parts of the country, and could be very difficult to reconcile with overall sustainable development goals. The Government will need to provide a framework in which people can exercise their transport choice in ways which are compatible with environmental goals.

In *Planning policy guidance: transport*, note 13 (PPG 13), published in March 1994 (Department of the Environment/Department of Transport 1994), the Government recognized the need for closer integration between land-use planning and transport. The stated aim of the guidance (p. 1) was to:

. . . ensure that local authorities carry out their land-use policies and transport programmes in ways which help to:

- reduce growth in the length and number of motorised journeys;
- encourage alternative means of travel which have less environmental impact; and hence
- reduce reliance on the private car.

In this way, local authorities will help meet the commitments in the Government's Sustainable Development Strategy to reduce the need to travel; influence the rate of traffic growth; and reduce the environmental impacts of transport overall. These policies will also make a significant contribution to the goal of improving urban quality and vitality, and to achieving a healthy rural economy and viable rural communities.

The document acknowledged that a high proportion of post-war development, particularly in the outer parts of conurbations, had been posited on the use of private transport—for example, the low-density housing estates with ramified streets, which public transport cannot easily serve, and the large out-of-town facilities such as hypermarkets and retail warehousing complexes, which depend upon and generate massive movements by car. John Gummer stated that he did not expect to authorize many more such developments if they came to him on planning appeal.

PPG 13 did not, however, offer the national transport strategy that the Countryside Commission and others had been calling for. In particular, it reaffirmed the previous official view that the trunk roads programme should be seen as a predetermined element in the planning process. That seriously weakened the capacity of the development plan system to integrate land-use and transport planning. Indeed, it appeared to pre-empt the essential debate about the acceptable limits on the capacity of the trunk road system.

The Department of Transport meanwhile proceeded to set up the Highways Agency, with a brief to pursue the massive programme of construction and improvement to major roads, despite the rising protests over the massive damage to the countryside caused by, for example, the great cutting that carries the M3 through Twyford Down. The minister gave way to the protestors by scrapping plans for the road through Oxleas Wood, but proceeded with his plans to introduce motorway charging, with the primary aim of raising money to build major roads faster.

THE SCENE TRANSFORMED

And yet, since mid-1994, the scene has been transformed. In Chapter 1 of this volume, Phil Goodwin speaks of the 'new realism' related to trunk roads and motorways. He records that the British Road Federation's report, published in summer 1994, proved what many of us had suspected—that a trunk roads programme of £2 billion a year (as it then officially was), and even one of £3 billion a year, could not keep pace with the traffic growth, on present policies and trends. In the 3 months between the delivery of Phil Goodwin's lecture and that on which this chapter is based—such is the speed with which the climate of thought is shifting in the new direction—two further seminal reports were published.

First, in October 1994, came the report *Transport and the environment* from the Royal Commission on Environmental Pollution (1994), which gives an authoritative description of the adverse effects of the present transport system, some of which I earlier described. The Royal Commission endorsed the framework for a sustainable transport policy as set out in *Sustainable development: the UK strategy* (Secretary of State for the Environment *et al.* 1994), and called for radical action to realize such a policy. This includes: halving the expenditure on motorways and trunk roads; a massive switch of resources into public transport; increased taxation on private transport to reflect the damage it does to health and to the environment; and much fuller integration of transport and land-use planning. Sir John Houghton,

Chairman of the Royal Commission, explains their conclusions more fully in Chapter 2 of this volume.

The second report was that of the Standing Advisory Committee on Trunk Road Assessment (SACTRA), published in December 1994 under the title *Trunk roads and the generation of traffic*. The purpose of this report was to address the question of whether new or improved trunk roads induce extra traffic. Until now the official assumption, built into the way that new road schemes were assessed, was that road building did not generate extra traffic and thus that, for example, a new road would be wholly used to deflect existing traffic away from the congested places that it was designed to relieve. Other people suspected that what you might call 'car Parkinson's law' would operate, with traffic expanding to fill the space provided for it.

The SACTRA report contains some sensible cautious language, but it concludes that:

- induced traffic can and does occur, probably quite extensively;
- the economic value of a new road scheme can be overestimated by the omission of even a small amount of induced traffic;
- there is therefore a need for a change in appraisal practice.

The government's response to SACTRA (Department of Transport 1994) was published at the same time as the report. It accepts the need for change in appraisal practice, and says that the likely significance of induced traffic will be assessed in relation to every new national road scheme. The government's response to the Royal Commission is likely to take a bit longer to appear. But the recent autumn budget contained a substantial scaling down of the trunk road programme.

IMPLICATIONS FOR THE COUNTRYSIDE

Thus, things appear to be moving in the right direction. But what does this imply for the countryside?

A rise in taxation does not stop people buying cars. It does not stop the car companies from advertising their wares—or from suggesting through their advertisements that the vehicles are either small enough to get into any city centre parking space, even though on-street parking may be banned in many of our city centres, or rugged enough to get to any mountain top, however illegal that may be! Similarly, a cut in the motorway and trunk road programme may mean, looking years ahead, that some stretches of countryside will be saved from blight and that less new traffic will be induced. But it does not, by itself, change the way in which people use their cars; indeed, until other measures are taken, it may simply prompt people to use, for their commuting and other journeys, roads through towns or villages that do not have the safe capacity for such use.

In this context, we need what I will call a fourth dimension to our new-look policy for road and road transport. The first three dimensions are already in, or coming into, place: *first*, traffic restraint, and growing emphasis upon public transport, in the cities; *second*, a more realistic and constrained approach to our trunk road programme, and a boost to long-distance public transport; *third*, a change in the climate of taxation, control of emissions, and the like in order to dampen demand and reduce environmental impact. The *fourth* dimension, inadequately treated in the report of the Royal Commission, is *strategic traffic management in and for the countryside*.

Traffic schemes in parts of the countryside are not a new idea. Traffic congestion has been a problem in certain areas since the 1930s; and serious efforts to tackle it began in the 1960s, notably in the National Parks, where recreational traffic was heavy and concern for the environment was high. Two examples from the Peak District will illustrate the approach.

In the Goyt and Derwent Valleys, the attraction of reservoirs and surrounding hill scenery was such that many thousands of people would arrive by car on a summer weekend: the narrow roads up these valleys became choked with traffic, giving no fun to those in the cars, spoiling the scenery, and impeding people who wanted to walk along the road. The National Park authority created car parks in the lower parts of the valleys, closed the roads at weekends, and provided minibus services for those who could not, or did not wish, to walk, and a cycle-hire service, which has proved to be extremely popular. The public response was very favourable; people enjoyed the quietness of walking or cycling along the roads, and, after a while, even the minibus services were withdrawn.

The other example is the 'Routes for People' scheme, which was designed to segregate traffic in the central part of the Peak National Park. Here, the main problem was the impact of heavy lorries from the limestone quarries, which were thundering through the villages and cutting up the tourist traffic along the narrow lanes. The National Park Board, working with the highway authorities, identified specific routes, running clear of the villages, that were suited to lorry traffic: these routes were widened where necessary, but kept in local character by a careful rebuilding of drystone walls. An agreement was reached with the quarry companies that their lorries would use these routes. Other roads were then marked as a 'White Peak scenic drive' for visitors.

These initiatives, plus similar action taken in other National Parks or scenic areas (such as the 40 mile/hour speed limit imposed throughout the New Forest), were reactions to crises of congestion or danger. They did not address the wider question that we now recognize of insidious damage to countryside.

The Royal Commission gives some sense of this broader issue in the following passage in their report (Royal Commission on Environmental Pollution 1994, p. 181).

> Perhaps the most intractable problems arise in the less isolated rural areas, especially in the vicinity of large conurbations. Here, road improvements, population dispersal and changes in land use have combined to produce large flows of traffic with consequently high levels of pollution and pressures for the construction of new infrastructure. (Traffic levels on major roads in Surrey, to take an extreme example, are already twice the national average). The scale and complexity of the problems in these areas will require the full range of solutions which are advocated in this report.

We in the Countryside Commission would argue that the 'full range of solutions' advocated by the Royal Commission on Environmental Pollution will be needed, in due course, in most parts of the wider countryside. Meanwhile, the reference to Surrey is very pointed for us, for this is one of two counties (the other being Cumbria) in which we are now working with the highway authorities to address precisely and extensively the problems to which the Royal Commission refers.

Before I describe our practical development work with these two counties, let me make a crucial point about public awareness. The motor vehicle, and particularly the private car, has brought such convenience, such pride, and such perceived freedom to move around that those who place disciplines upon its use can be very unpopular. Such disciplines will not work unless they have popular acceptance. Thus, the traffic management in the Goyt and Derwent valleys of the Peak District responded to frustration among visitors and local farmers. The 'Routes for People' scheme was prompted by protest among villagers about quarry traffic. The New Forest speed restriction was imposed after public outcry over ponies killed by fast cars.

What is encouraging about our initiative with the county highway authority in Surrey is that it follows a growing clamour among residents of that county to reduce the impact upon their quiet lanes and leafy housing areas of the 'rat-run' of commuters and other traffic, which has been growing year by year. The residents' associations of Surrey were pressing the County

Surveyor to tackle this, and some were even offering to pay for traffic-calming schemes themselves. These were people who themselves own and use cars, usually more than one per household. Just as a pedestrian in the town centre can be defined as a motorist who has found space to park his car, so the rat-run commuter, when he gets home, becomes the resident who wants his quiet to be undisturbed by traffic!

This is part of a broader frustration about the environment, health, accessibility of services, and the general quality of life in such areas. It is clear that we are reaching the point of break-through when individuals accept the need for collective discipline on, for example, the use of cars. The recent change of tone within the RAC and AA on this subject is striking.

The initiative in Surrey was launched in 1993, under the title STAR (Strategic Traffic Action in Rural Areas). It is jointly run by Surrey County Council and the Countryside Commission. The aim is to sustain and enhance the quality of the countryside by examining the impact of traffic on it and applying the principles of demand management, taking one step at a time through careful local consultation. An outline of strategic ideas was produced and was the subject of extensive consultation with parish councils, residents' associations, district councils, and other organizations. The aims were widely endorsed, and an initial set of five local schemes is now being designed to contribute to the long-term strategy. These schemes, briefly, focus upon:

(1) the Dorking area, where 'rat-running' traffic is causing noise, pollution, fear, and intimidation: the aim will be to encourage people to use roads in accordance with the county's new strategic road hierarchy and thus to remove the rat-running traffic from where it should not be;

(2) the village of Shere and its attractive surroundings: the aim will be to reduce the volume of visitors' traffic, thus improving the natural environment and at the same time enhancing the local economy;

(3) a country cycling network: the aim will be to increase the safety and pleasure of cycling, by designing and promoting cycle routes and educating all road users about living together if total separation of drivers and cyclists cannot be achieved;

(4) south-west Waverley, a beautiful part of the Surrey Hills, including extensive National Trust lands: here the aim would be to create a much safer and more tranquil area, mainly for walkers, horse-riders, and cyclists;

(5) a schools scheme, with the aim of getting children and their parents to address the effects of using cars and to work out agreed alternatives, such as car-sharing schemes or pick-up buses.

These are the first pieces of an exciting strategic jigsaw in the whole county of Surrey, outside of the M25 ring. We are grateful to the County Council for being so keen and resourceful in working with us to find the strategic solutions. With them and others, we hope we can build quickly a network of good practice that will transform the prospects, willed by the public but so difficult to articulate and apply, of sustaining the quality of the English countryside.

The work in *Cumbria* is focused on a new roads hierarchy for the whole of the Lake District National Park. The aim is to develop dedicated access networks for different classes of road users, such as cars, buses, coaches, walkers, and cyclists. These networks will be worked out on a local basis, building up over time (and as money becomes available) to form a park-wide pattern. Again a set of initial projects is being prepared for extensive consultation, which will:

(1) use agreed traffic signing and speed restrictions to enable traffic to remain on main traffic routes instead of spilling on to minor roads;

(2) use agreed parking and access arrangements to alleviate congestion in popular villages;

(3) give walkers and cyclists the dominant use of a network of minor roads, by prohibiting inessential access without damaging local interests;

(4) increase the convenience of public transport wherever possible;

(5) use traffic-calming measures on sensitive areas like commons, in order to increase safety for local people, livestock, and visitors.

We in the Countryside Commission are also supporting programmes of traffic management in other places, notably Dartmoor and North York Moors (with the RAC). We hope that these and other initiatives will reduce the adverse impact of roads and road traffic upon these areas and produce ideas that others can use elsewhere. Further advice, based on the work we have done so far, is given in an advisory booklet (Countryside Commission 1996).

There are two other areas of concern to the Countryside Commission, related to planning and design of roads. I wrote earlier that the national debate has moved towards challenging the concept that we need yet more main roads. Indeed, I believe we should ask, 'Are these journeys, these new roads, really needed?' To probe this question properly, we need a system of proposal and appraisal that permits full debate about the totality of a major route, such as the Harwich–Fishguard trunk road or the major road along the south coast of England, rather than a piecemeal process of considering limited sections of such routes. Ideally, such debate should be set in the context of a national transport strategy, and regional planning frameworks. It should address the overall need for the road, bearing in mind (what is now recognized) that a new or much improved route may well induce new traffic; and it should also consider the broad alternative alignments for the road, and the environmental and other impacts.

From the standpoint of the countryside, I recognize that, within a new and more integrated transport policy, a sound case may well be made for some new main roads or road improvements, particularly in order to remove through traffic from towns and other eroded or sensitive places. The new approach must therefore include thorough-going environmental impact assessment and a creative approach to enhancing the 'view from the road'.

There is growing public and political recognition that the impact that major projects would have on the environment, broadly interpreted, should be fully assessed as part of the process of deciding whether they should proceed. The European Commission is moving, indeed, towards acceptance that not only projects, but also programmes and policies, should be subject to such assessment. At present, the major road programme in this country falls significantly short of this intention.

The case for a new widened trunk road is made largely on grounds of traffic volumes, travel time, and road safety. The changes in appraisal procedure following the SACTRA report (Department of Transport 1994) may tilt the case against some schemes that would otherwise have proceeded. But questions of environmental impact ought, in our view, to be given greater weight than they now receive.

Environmental assessment tends to be undertaken only at the stage of selecting the detailed route, and on the basis of procedures set out in the Department of Transport's (1993) *Design manual for roads and bridges* (formerly titled *Manual of environmental appraisal*). These do not, in our view, adequately appraise the environmental costs and benefits of a proposal, nor is the appraisal integral to the basic decision as to whether the project should be brought forward. In parts, the *Design manual for roads and bridges* is more sensitive than its predecessor in describing the impact of road schemes—opening up, for example, the potential but not yet the common practice of allowing assessment of 'the combined and cumulative impacts of several schemes'. But we believe that a far more searching process of environmental impact assessment is needed, especially at the stage of policies and programmes rather than that of individual projects. The *Design manual for roads and bridges*, republished as recently as June 1993, appears to preclude this.

As to the view from the road, the Countryside Commission have welcomed the concern and expenditure that the Department of Transport has committed to the landscaping of motorways. The alignment, elevation, landform, detailing, and landscaping of some of these and other major roads has enhanced the quality of the journey and, in places, opened new views of spectacular or attractive countryside. But the so-called 'improvement' of roads, and the street furniture and other features that accompany road traffic, still too often do grave damage to the beauty and character of the countryside.

The Countryside Commission's (1995) advisory booklet provides practical guidance on best practice in the detailed assessment of new road proposals and in the landscape treatment of new rural roads and the maintenance and management of existing ones. This advice will relate mainly to non-trunk roads, but many of the principles are also applicable to trunk roads.

CONCLUSION

I conclude by quoting a comment made by the Steering Group in the Ministry of Transport that, 30 years ago, oversaw the production of the report *Traffic in towns*. This comment (Ministry of Transport 1963, intro., para. 53) graphically describes the challenge that we still face today—how to gain the benefits, while avoiding the adverse impacts, of the motor vehicle:

> We are nourishing at immense cost a monster of great potential destructiveness. And yet we love him dearly. Regarded in its collective aspect as 'the traffic problem', the motor car is clearly a menace which can spoil our civilisation. But translated into terms of the particular vehicle that stands in our garage or outside our door, we regard it as one of our most treasured possessions or dearest ambitions, an immense convenience, an expander of the dimensions of life, an instrument of emancipation, a symbol of the modern age. To refuse to accept the challenge it presents would be an act of defeatism.

I believe that we now have such public acceptance of this dilemma that we can indeed accept that challenge. People want the benefits of the motor vehicle, but they also want to sustain the quality of their lives and the quality of our countryside. Strategic traffic management offers the means of reconciling these two desires. We are still at the stage of testing out how it should be done. It will not be cheaply or quickly done. Wider programmes of public information and education will be needed. But the new realism has arrived and is gaining ever wider acceptance among the people and within government.

We in the Countryside Commission, with our partners, will press forward with the work we are doing to apply that new realism in the countryside, so that the qualities that people have for so long valued in that countryside are not diminished.

REFERENCES

Countryside Commission (1992). *Road traffic in the countryside*. Countryside Commission, Cheltenham.

Countryside Commission (1995). *Roads in the countryside*. Countryside Commission, Cheltenham.

Countryside Commission (1996, in press). *Countryside traffic action* (provisional title). Countryside Commission, Cheltenham.

CPRE (Council for the Protection of Rural England)/Countryside Commission (1993). *Tranquil areas: south-east England.* CPRE/Countryside Commission, London.

Department of the Environment/Department of Transport (1994). *Planning policy guidance: transport* (PPG 13). HMSO, London.

Department of Transport (1989). *National road traffic forecasts (Great Britain) 1989*. HMSO, London.

Department of Transport (1993). *Design manual for roads and bridges.* Vol. II, *Environmental assessment*. HMSO, London.

Department of Transport (1994). *The Government's response to the SACTRA report on 'Trunk roads and the generation of traffic'*. HMSO, London.

Grahame, K. (1908). *The Wind in the Willows*. Methuen, London.

Jeffrey, R. (1949). *The King's highway*. Batchworth Press, London.

Ministry of Transport (1963). *Traffic in towns: reports of the Steering Group and Working Group appointed by the Minister of Transport*. HMSO, London.

Oxford University Transport Studies Unit (1992). *Trends in transport and the countryside.* Countryside Commission, Cheltenham.

Royal Commission on Environmental Pollution (1994). *Transport and the environment*, 18th report of the Royal Commission on Environmental Pollution. HMSO, London.

SACTRA (Standing Advisory Committee on Trunk Road Assessment) (1994). *Trunk roads and the generation of traffic*. HMSO, London.

Secretary of State for the Environment, Secretary of State for Foreign and Commonwealth Affairs, Chancellor of the Exchequer, President of the Board of Trade, Secretary of State for Transport, Secretary of State for Defence, *et al.* (1990). *This common inheritance: Britain's environmental strategy*. HMSO, London.

Secretary of State for the Environment, Secretary of State for Foreign and Commonwealth Affairs, Chancellor of the Exchequer, President of the Board of Trade, Secretary of State for Transport, Secretary of State for Defence, *et al.* (1993). *UK strategy for sustainable development: consultation paper*. HMSO, London.

Secretary of State for the Environment, Secretary of State for Foreign and Commonwealth Affairs, Chancellor of the Exchequer, President of the Board of Trade, Secretary of State for Transport, Secretary of State for Defence, *et al.* (1994). *Sustainable development: the UK strategy.* HMSO, London.

5

Railways and sustainable development

Bob Reid

Sir Bob Reid took his first degree at the University of St. Andrew's and joined the Shell Oil Company in 1956. He served the company for 35 years, in Brunei, Nigeria, Kenya, Thailand, Australia, and in the London headquarters. In 1985, he was appointed Chairman and Chief Executive of Shell UK, a post that he held for 5 years before agreeing, in 1990, to become Chairman of the British Railways Board. Sir Bob left British Rail in 1995 and his Linacre Lecture was one of his last major public statements as head of Britain's railway industry. He is now Chairman of London Electricity plc and of Sears plc.

INTRODUCTION

There is an old Chinese adage that says 'In action, watch the timing'. Obviously Linacre College is well aware of that message and has been a complete master of the art in choosing the right time for the series of lectures that form the basis for this volume.

There could hardly be a more topical subject than transport and the environment. Until relatively recently, each issue was something of a Cinderella, required to leave the political ballroom just when the serious fun was supposed to start. But for some years now, both have been gradually—and separately—moving nearer to centre stage. And 1994 may well go down in history as the year when it was fully recognized that they are inextricably interlinked and that successful management of them is of profound significance for this and future generations of Britain's citizens.

At the time I was preparing the lecture on which this chapter is based, the report of the Royal Commission on Environmental Pollution (1994) had not been published. However, it seemed likely that it would reinforce the main thrust of what Phil Goodwin said in his excellent first lecture in the series (Chapter 1, this volume). So I have in any event taken that as the backdrop to my comments, and below I summarize briefly the relevant points of his argument.

First, Dr Goodwin gave the definition of 'sustainable development' as (p. 6, this volume) 'meeting the needs of the present without compromising the ability of future generations to meet their own needs'. Others have suggested more stringent standards that seek to ensure that resources —renewable or non-renewable—remain available and that pollutants can be assimilated. But the definition taken by Dr Goodwin will certainly do. We all know essentially what we are aiming for.

Dr Goodwin went on to describe the main components of congestion, pollution, and damage to the environment and to health caused by the unchecked growth in traffic, particularly on the roads. He explained how there is now common recognition as far as urban transport is concerned that we cannot continue with policies of building roads to meet forecast demand, that demand management has to be a part of transport policy, and that, among other things, public transport needs to be improved and extended.

Dr Goodwin further argued that implementation is lagging behind the acceptance of policy in relation to urban transport and that policy thinking about areas outside our towns and cities is some way behind that for urban transport, but that similar recognition is emerging there too.

Against that background, I propose to examine:

- how the railways can contribute specifically to the aims of sustainable development;
- what the industry has to do to attract more people to use the services it has to offer;
- the policy challenge the present position represents.

WHAT CAN THE RAILWAYS DO?

The first issue in relation to the railways is the extent to which they can help replace road as a means of transport. What can they contribute to solving the problems of producing sustainable transport policies? Are they in fact important enough to make a significant difference, bearing in mind that rail is now responsible for under 10 per cent of both passenger and freight traffic?

Here the answer is a very definite yes. Rail can be particularly effective in areas where the problems of congestion and pollution are most acute. Of course, rail does not compete for the millions of everyday short journeys that people make or for the work of local delivery vans and smaller lorries. But it has a key role in commuting. In London, the railway system and London Underground together satisfy some 70 per cent of commuters' needs. Elsewhere the position varies but rail has the potential to do much more than it does.

InterCity services are the single most important carrier for a number of routes between London and major centres in Britain, thus releasing pressure on many of the country's key arterial highways. There is scope for winning yet more business and reducing the load that would otherwise fall on the motorways and trunk roads.

On the longer freight journeys—those over 200 kilometres—rail is a serious challenger to the heavy lorry. It is often pointed out that a doubling of rail's share of the freight market as a whole would only reduce road's share by some 5 per cent. But, looking specifically at this heavy long-distance road traffic, doubling its share could reduce the need for the very heavy vehicles by some 45 per cent—a very different picture from the comparison with road freight as a whole.

Energy use and pollution

But, if the shift from road to rail is to help with sustainable development, rail must demonstrate its 'green' credentials. What are they? Most directly, they stem from the effective use of energy, that is to say, energy use in the motive sense—taking people and goods from A to B.

The starting point of rail's efficient use of energy is the inherent efficiency of a steel wheel rolling on a steel rail with the associated low levels of friction. That same characteristic causes an operational problem—trains are not very good at climbing hills. To get over that, the railway builders have to construct tunnels, cuttings, and viaducts to avoid anything approaching steep gradients. Thus, when lines are built, there can be the fiercest of arguments about environmental damage and loss of amenity from the construction itself—and about the cost of mitigation. That has not been an issue in Britain for the best part of this century, although it has recently arisen in the case of the planned new Channel Tunnel Rail Link where the attention paid to environmental factors has been immense. Essentially, however, the railway in use in Britain today was built before the century began. As a result, we, as inheritors of the system, can enjoy the added advantage of the lower amount of energy required to get past the hills and valleys that have, in effect, been flattened out for the railway user.

The railway can carry people and goods in much greater bulk than is possible on the roads. The fact that the traffic can be—and indeed has to be—directly controlled on the permanent way means that potential congestion can be managed. Wasted time and energy at peak times are thus reduced to a minimum, and the train's dedicated right of way gives it faster and more direct access than that of any other mode of transport to the hearts of most of the country's major towns and cities. The railways cause noise and have to take steps to minimize the nuisance thus created but, in general, they are less intrusive than roads.

Another feature of the railways has been their ability to make use of different sources of tractive power, since they are not tied either to one type of primary fuel or one method of combustion. The shift away from coal and steam power was made for economic, not environmental reasons. But, even if the economics had remained favourable to coal, it is hard to conceive how a coal-based railway network could possibly have survived the growing environmental concerns of the past 40 to 50 years. Note that, in the first year of the life of the nationalized railway in Britain, 20 000 steam locomotives burnt 14 million tons of coal. And their thermal efficiency was only some 5–7 per cent compared with around 35 per cent for coal-fired power stations.

Today's railway is run with a combination of diesel and electric power. While around one-third of the network is electrified, electric power accounts for half the country's train mileage. That helps remove fumes and pollutants from crowded urban areas. It also means that, for half our operation, the extent of gaseous emissions depends on what the electricity generators do. The 'dash-for-gas', as gas-fuelled combined cycle plants displace the coal-fired stations in the hierarchy of electricity generation, will mean a further reduction in the level of gaseous emissions attributable to the railways.

Those are the underlying factors that lead, on average, to a passenger-kilometre by train accounting for the consumption of 40 per cent less energy than one by car and a tonne-kilometre of railfreight for 80 per cent less energy than if the freight were moved by road. The corollary is fewer emissions, particularly of carbon dioxide, CO_2.

It is, however, critical to recognize the effect of load factor. Full trains can be greatly more fuel-efficient than cars with no passengers or lightly loaded lorries. But empty or near-empty trains change the picture completely. Plenty of analyses have been made of the relative impacts and I will not consider them here. What is important to recognize is that, if you are going to assess responsibly the contribution rail can make, you have to take into account the different loadings the industry can achieve in the different operations it undertakes.

To conclude this section, a brief word on safety. I take it that sustainable transport has to put a premium on the safety of the people who use it. Every accident on the railways reminds us that we still have much work to do to ensure the safety we want. The tragedy at Cowden on 15 October 1994 was all the more disappointing as it ended a period of some 3 years during which no passengers died as a result of a train accident apart from one case caused by a wicked act of vandalism. We also mourned the death of three of our employees. But, in spite of such disasters, rail remains much safer than the car and thus offers the nation's travelling public a further advantage that needs to be weighed in seeking a rational balance between modes of transport.

Use of resources

So far, I have outlined the bare bones of rail's credentials for helping to find the answers to the problems of congestion and pollution. There is also the question of the wider use of resources—including the energy put into creating the assets that the industry uses.

As far as I am aware, there does not exist any attempt at a full input–output analysis, comparing the total resource use of the different modes of transport. Thus, we cannot make a full comparison. But the railways do make use of assets with very long lives. One bit of arithmetic we did some time ago showed that our programme of replacing bridges implied an average life for each of them of some 800 years! They may or may not survive that long, but I cannot imagine that any such calculation would normally figure in the basic assumptions for even a supremely optimistic company's corporate plan.

Going back to the earliest railway days, what was created was a network for freight that was also used by a few passengers. Today's railway in Britain is very different. It is essentially a system for passengers with some very important but much less numerous freight flows. Within the passenger businesses are services with very different characteristics—high-speed intercity, commuting, cross-country, and rural. There are very heavy bulk freight loads and very different flows of container traffic. It is about as complex a mixed railway as you will find anywhere.

And yet it is essentially based on the network that was laid down in the Victorian era, 100–150 years ago. If sustainable development involves making the best use of the assets you have rather than scrapping them and using more resources to replace them, then the railways can claim an impressive record. A lot has had to happen to make that possible.

Particularly over the last 40 years or so, the industry has increasingly had to adapt to compete with other forms of transport. Thus it has had to change to meet modern needs. The outstanding example is the story of the development of Britain's high-speed passenger network. The ability to run a network of fast, high-quality, frequent intercity services throughout the country on existing infrastructure is recognized around the railway world as a particular British genius. We do not achieve the very high speeds of the Japanese, the Germans, or the French. But, given our density of population, our terrain, and the distance between main population centres, a very high speed system dependent on dedicated new tracks is not best suited to what we need.

Dr Goodwin (Chapter 1, this volume) writes of the vicious circle in which road improvements attracted people away from public transport, which then found it increasingly difficult to compete. The story of the development of intercity services is one of a virtuous spiral going the other way. Diesel and electric traction, replacing steam in the early 1960s, made for faster journeys; people were attracted to the service—notably a 33 per cent increase in 2 years after the high-speed trains were brought into the Great Western Line services to Bristol and Cardiff in 1976; quality and frequency increased; market pricing was introduced; and performance justified further investment. The most recent component was the upgrading of the East Coast Main Line and the introduction of the InterCity 225 service. I look forward to seeing as soon as possible the upgrading of the West Coast Main Line to take the process further.

Similar ingenuity has been shown in adapting the infrastructure to meet modern-day commuting needs. Refined logistical planning in London has been so successful that the number of peak-time trains has actually exceeded the design capacity of some London termini. Thameslink—after 6 years of operation now carrying some 450 million passenger miles annually—was the result of bringing back into use a section of track that had not been used for passenger trains for some 72 years.

Such a network, essentially sharing track with the rest of the railway, could not work without advanced control and signalling systems. To give an example of how 'mixed' a railway it can be, there are some 400 trains that pass through York station on an average weekday. Fifty-four are high-speed InterCity 225s, 24 are InterCity cross-country, over 100 are other express services, 71 are local passenger trains, and there are approximately 130 timetabled freight trains every 24 hours, half of which travel in daylight hours. To cope with such problems, the integrated electronic control centre was developed—a technology in which Britain is an acknowledged world leader.

The story has been one of continual adaptation to take advantage of technological and market opportunities as they arose in the face of almost overwhelming competition from road and air. The result has been the development of a network that is essentially cost-effective and fit for the purposes and the markets it has been designed to serve. The above are simply some examples of what it has taken to turn our inheritance of a Victorian infrastructure into the basis of an effective, modern railway.

Our rolling stock too is a relatively long-life asset. A lifetime of 30–35 years is standard, although there are not many 30–35-year-old road vehicles in regular use today. While good in terms of sustainable development and the use of resources, the long life of the rolling stock does present us with commercial problems. A 30-year-old carriage is perfectly serviceable but it does not give the passenger much of an impression of a lively up-to-date business.

As with the infrastructure, there are many examples of how resources have been used increasingly effectively while extending and improving the services that the railway offers.

1. When the industry was nationalized, the average load factor per freight train was only half of today's figure, and the number of passenger miles carried by each coach was a fifth of that achieved today.
2. Travelling from here to London only some 5 years ago, passengers would probably have been on a nine-coach train, weighing 275 tonnes, hauled by a 100-tonne locomotive with huge fuel consumption and the train would have been only one-quarter full. That journey would now be in an air-conditioned class 166 train, weighing only 115 tonnes, carrying more passengers, and using less energy.
3. On the electrified commuter railway in the south-east, the new Networker trains carry 15 per cent more passengers using 25 per cent less energy than the trains they replace. Aluminium bodies and alternating-current traction motors with more efficient control equipment and regenerative braking have contributed here.
4. On the East Coast Main Line, the margin for maintenance and to cover the unexpected is only 16 per cent of the vehicle fleet—far less than for other high-speed networks.
5. Disc brakes are hugely more cost-effective than their cast iron predecessors. They last much longer, are not so noisy, and do not spread iron filings liberally over the countryside.
6. Improvement in productivity in the use of assets has intensified. In 1993 the industry carried about the same level of passenger miles as 10 years ago with around two-thirds of the passenger coach fleet—a 45 per cent improvement. The productivity of the use of freight wagons has shown even more impressive increases.

Thus, the railways can help to provide journeys with relatively efficient use of energy and reduced noxious emissions, the industry has contributed to sustainable development by seeing to it that assets passed on to us from the last century can be used to good effect, its mobile assets last considerably longer than those of most of our main competitors, and it has shown massive improvements in the efficiency with which those mobile assets are used.

Most of the examples I have given involve cases where the two factors, commercial aims and the requirements of sustainable development, reinforce each other in things we can do. This includes propositions such as 'lower energy use means lower costs, which help to attract business'. The two factors can also combine to tell us what we should *not* do. A prime example of that is found in the field of freight. There is almost universal support for moving freight from road to rail. In principle, we, of course, would like that to happen too. But not at any price.

The price of Speedlink—our wagonload freight service—was too high. Financially it was a disaster—losing us some £30 million a year on a £45 million turnover. But, in terms of energy use and emissions, it was also grossly unattractive—massive, heavy diesel engines had to travel miles with numerous small payloads in order to marshal together full trainloads. Thus the case for abandoning Speedlink was clear, both on business and environmental grounds—particularly as there were prospects of retaining much of the traffic by reorganizing arrangements to provide trainload quantities of freight.

Capacity

Having, I hope, essentially set the scene, I come to what I see as the nub of the issue. It is the simple question of capacity.

Dr Goodwin's thesis is that more roads cannot provide the capacity needed to cater for unchecked growth in demand. Rail certainly has the capacity to spare, although this is not absolutely universally the case. Many peak-hour commuters on overcrowded trains would testify to this, as would those packed into the InterCity holiday weekend favourites or trains going to some of the prime tourist targets. There are also some infrastructure bottlenecks. For example, the whole justification for the new Channel Tunnel Rail Link is the prospective shortfall in capacity in the south-east, and projects like Crossrail are there to help deal with central London congestion.

Nevertheless, across the country there is great scope for trains to carry many more passengers at little or no incremental cost. Across the network as a whole, the average number of people on each passenger train service we run is 83. While we compare well with our European counterparts on measures such as the costs of putting trains on the track and running them, passenger loading is only around 60 per cent of the European average.

This does not necessarily mean we are less efficient. The nature and pattern of the demand can be very different. Even with the busy London commuter services, the recession has hit commuting numbers so that we are carrying some 20 per cent fewer people into London at peak times than we were 5 years ago. Compared with our European counterparts, we tend to run more frequent services of shorter trains and 83 on a two-car train is a very satisfactory number. And, of course, higher subsidies on other major European railways often mean much lower prices for the passenger. Nevertheless, the figure represents only some 25 per cent of the capacity we provide, showing up the massive potential we have for increasing passenger numbers without either extra cost or noticeable extra use of energy or other resources. This is true even before any calculation is made of the impact of simply increasing the number of coaches on trains let alone of modifying the infrastructure. Thus the question here is how do we attract more people to use the railways.

HOW CAN THE RAILWAYS ATTRACT MORE CUSTOMERS?

Let me digress and mention a painting in the dining room in Euston House entitled 'Going north'. This is a large picture probably painted in the early part of this century. There are many groups of people preparing to join a train: parties going to shoot Scottish birds with guns and dogs, looking rich and ready for the moors, at least for the first mile; some people going to fish, looking more intellectual than rich; policemen addressing and arresting a likely criminal; children running around everwhere. All beginning an exciting and attractive journey. These were our customers. Why? Because they had no alternative or because the whole experience was exciting.

As I look at the crowds joining the Eurostar services, I am reminded of that picture. Admittedly, Eurostar is closer to its alternatives and it faces more competition. But I certainly see a lot of the tingling excitement that the painter of 'Going north' perceived in that earlier age.

My mind goes to another picture, a mural that adorns the wall of the foyer of the Shell cinema at Waterloo. This mural is all about Londoners and the war. Buses, planes, and people scurrying everywhere. What strikes you when you look carefully is that all of the many people that make up the mural are wearing hats. Draw that mural today and you would be lucky to find 10 hats. Fashions change and times change. In symbolic terms, the problem of the railways is to put the hats back on their public. Not to take them back in time but to rekindle the idea that rail travel is the natural and most congenial choice for many of their journeys.

We know we can attract people to rail under the right circumstances. We have been able to put hats back on some of the people as the following examples indicate.

- Attractive marketing packages, with Apex and Saver fares and imaginative promotions, have increased load factors both on InterCity and other services. One of the notable features of some campaigns has been the number of people who were enticed back to the railway who had either hardly used trains at all before or had not done so for many years.
- Total modernization of the Chiltern Line has turned a route that was seriously considered for closure only a few years ago into a popular line with 25 per cent more passengers than it had before, many of them drawn from the M40.
- On the Manchester–Leeds trans-Pennine route, the introduction of new, air-conditioned trains, running every 20 minutes instead of once an hour, has increased demand by some 60 per cent—and at least some of the demand will have been taken from the busy M62.

There are many other examples of similar initiatives elsewhere. Much is being achieved—a lot of it in fruitful co-operation with local authorities and passenger transport executives (PTEs). New lines are being opened, new services introduced, and new stations opened—some 250 new stations over the last 15 years.

The effort needs to be continued and intensified. There is plenty of off-peak capacity to be filled. Can we attract more shoppers to follow commuters into city centres when the commuting peak has died down? Can we create more of a 'return load' for commuter trains, picking up people who want to travel into the major urban centres which are the start of many commuting journeys? If we are to be successful at that sort of thing, we may need to be more imaginative than we have been to date in linking in with other modes of transport. This is a point of some importance.

For freight, the 'intermodal' concept is well established. All see that rail's strength is in the trunk section of the journey not the door-to-door collection and delivery. Piggy-back and swap-body techniques are continually advancing opportunities. Specialist freight forwarders have emerged who can secure value for money by choosing the best combination of modes for each client.

For the passenger railway, intermodality has been seen in examples such as rail links with major airports. The link with Gatwick has been followed by those with Birmingham, Stansted, and now Manchester airports, and, although the collapse of the tunnel planned at Heathrow was a serious setback to the project, there is no doubt that the Heathrow Express will make an important contribution towards giving London the sort of transport system that a major financial and tourist capital needs for the twenty-first century.

Obviously, providing car parks at stations has for a long time helped to turn car-users into rail passengers—I wonder how many know that the combined amount of parking space at the stations of the former Network Southeast business is greater than that of the whole of central London? Park-and-ride schemes can link rail with public road transport.

But I have a feeling we may so far have only scratched the surface of what could be possible. We already offer car hire facilities at some stations—the obverse of 'park and ride'—and, if competition law allows, we might perhaps see the development of 'rail–drive' packages based on this. Could we envisage, say, battery-powered runabouts for those city-centre trips that have to be made at the end of a main-line journey by rail?

Perhaps the most important link is with the bus—or, in the light of its growing reappearance, the tram. Again there are some problems with competition law, particularly following bus deregulation. But, increasingly, people are recognizing that bus and rail services can combine to offer an alternative to the car that can only benefit the public transport user.

Then there is the bicycle. I am conscious that the railways have come to be seen as less friendly to the cyclist than in the past. Primarily that stems from our having rolling stock that is no longer well-equipped to carry bicycles—the prime purpose of a passenger carriage being to provide the passenger with a comfortable, problem-free journey. We cannot, in practice, redesign all our rolling stock for what is bound to be a minority activity. Perhaps, however, there will be more that can be done to provide for parking and storing cycles at the start of a rail journey with new initiatives for hiring them at the other end.

This is only conjecture and in no sense the result of detailed studies and costings. All I wish to do is raise the awareness that intermodality is the way in which the community we serve can be extended beyond those near to our network of 2500 stations. It could be win–win for passengers, as it clearly can be for the freight customer.

In summary, to maximize its contribution to a sustainable transport system the industry must continue in all the constructive directions I have outlined above:

1. We must ensure that our equipment and services are fit-for-purpose for the markets they are aiming to capture.
2. We must use technology, analytical skills, and plain sound management to ensure that all resources are used effectively and costs kept down.
3. We must continue research both to improve efficiency and to make the industry's activities ever safer and less damaging to the environment.
4. Attracting more custom to make use of the extensive spare capacity that exists on the network today will be achieved by:
   (a) focusing firmly on what the customer wants, concentrating all the efforts of the organization towards satisfying customers, and looking for continuous improvement in those efforts at every point.
   (b) paying attention to detail in all our production processes, using up-to-date methods of analysis to find out what people genuinely want from their journeys, showing liveliness in promoting our wares to the public at large, and putting an emphasis on service and customer-care in what we produce at the end of the day.

In short, we have to be a quality organization using quality processes to produce a quality product.

THE POLICY CHALLENGE

I have outlined what I see as the attributes of rail that enable it to help create a sustainable transport system, and what the industry needs to do to attract more people to use its services. I now turn to my third and final subject—the policy challenge.

The issue of sustainable transport represents a massive challenge for the political process. A number of broad policy areas are involved: economic; social; environmental; and fiscal. To the extent that travel and the carriage of freight is international, European and foreign policy come into it as well.

Demand management means some restrictions—and applying restrictions to well-established and entrenched behaviour across the population is a daunting exercise at any time. Increases in transport costs for better environmental protection raise arguments about maintaining our international competitiveness, and balancing those arguments with responsibilities for the general health of our nation's citizens and the well-being of future generations will call for the coolest and clearest of heads. In short, managing the process will demand political skills of the highest order.

Even given that the directions and aims of policy are clearly agreed, there is no guarantee of success in implementation. Even with a vision of the end objective the transition will be intensely testing, because, at root, what we have is a clash of freedoms—and clashes of freedoms are the very essence of political and social conflict.

Freedoms

Particularly in the heat of argument, freedoms are liable to be seen as absolutes. That is rarely the case. Freedom can be abused, misused, or taken to unacceptable extremes. I cannot think of a case where there do not have to be limits. For example, battles have been fought over many centuries over the freedom of speech, particularly as allied to the freedom of conscience and belief. Freedom of speech can be abused. It clashes with the freedom of individuals to have their privacy respected and to live their lives without hatred being stirred up against them on grounds of their race or colour.

In the twentieth century, freedom of mobility has become an issue. The citizens of the end of the Victorian era could hardly have imagined what we take for granted, namely, the opportunities there are for travel nationally and internationally. The right of the motorist to take his car when and where he likes has become so established that it would not be surprising to see it as a candidate in some people's minds for inclusion in any bill of rights. The *Daily Telegraph* of 19 October 1994 summed up its comments on the debate by saying: 'for all its undoubted disadvantages, the car confers a freedom of movement that most people take for granted. In future that may well have to be moderated, but it cannot be taken away. Any Transport Secretary who forgets that, does so at his peril.' Freedom of mobility has undoubtedly made our quality of life better. But, taken to extremes, it clashes with another freedom—the right to enjoy a reasonable environment in which to live.

Can anyone doubt that we have gone too far in allowing unfettered use of the internal combustion engine? What use is a freedom that simply leaves us stuck in tailbacks and traffic jams? Some 40 years ago the Clean Air Act cleared smog away from major cities. Should we today for a moment tolerate having to hear increasingly regular warnings to the elderly and asthmatic not to go out of doors because the air is not good enough for them to breathe safely? I think not.

The environmental case too can be taken to extremes and the argument misused. Practically any industrial development involves some detrimental impact on some part of the environment. Extremes of unthinking environmental protection would mean surrender of economic development and progress. That might be acceptable to those who are fortunate enough to lead comfortable lives today but is in no way acceptable to those who are further down the chain of prosperity, let alone those who suffer poverty and deprivation. If we raise our eyes from the problems of Britain and the Western world to the global picture, the next century looks set for a struggle between environmental and economic freedoms that could, literally, be a matter of life and death for the planet.

Thus, for Britain at the end of the twentieth century, the question of transport and the environment comes down to a balance between the freedoms the two represent. For sustainable development in transport what we are seeking is the protection of mobility without the environmental damage and distress that misuse of that freedom can cause.

This freedom had its first challenge in 1974. The first oil crisis revealed the emotions that bind motorists to their cars. Shootings in petrol queues surprised everyone. For the first time citizens and legislators alike took stock of where the motor car had taken them and they were astonished. Dependence on the Middle East changed the political dynamics of Western foreign policy. An anxiety to cover this vulnerability grew in the halls of Westminster and the capitals of Europe. Renewable resources became a high priority—restructuring of transportation modes received a good hearing and the recasting of work became a popular theme. Only one factor, however, made any real impact and then only a momentary one—price. Price rises make a dent in consumption, and higher prices highlight the importance of energy conservation. But the impact does not necessarily last. Cars become lighter forever, but only smaller for a short while.

The second oil crisis saw the 1974 process repeated but this time the effects were even more short-lived. The habit of personal mobility was too deep-seated to be shifted by a temporary disruption. The long line of single-person cars commuting into major cities bears continuing testimony to the phenomenon.

The resistance of the motorist to price increases immediately sharpens up the problems that any policy that seeks to dampen down car use will face.

The economic and social problems of forcing up petrol prices high enough to have a significant impact on demand could not be more obvious.

This illustrates the need for both sticks and carrots in any policy. The key measures already taken or under review by the government show examples of both: explicit acknowledgement that road capacity cannot be increased to cater for forecast demand; increasing restraint in urban areas; the commitment to substantial real increases in petrol tax; motorway tolls and urban road pricing; but also the promotion of public transport through the integration of transport and land-use policies advocated by the Department of the Environment/Department of Transport (1994) in *Planning policy guidance: transport* (PPG 13). Even though the land-use horse has got at least halfway across the paddock, but not to the stable door, it has somewhat surprised me that more recognition has not generally been given to PPG 13 and its potential significance for the longer-term development of urban transport in Britain. Even if practical changes on the ground as a result of PPG 13 are inevitably some way off, as an indication of the new direction in government thinking it could hardly have been clearer.

Those measures are entirely in line with Dr Goodwin's analysis of policy trends. As he has indicated, the full framework of relevant policies has still to be worked out. What are the principles that should underlie this framework?

External costs

First, social costs—internal and external—should be reflected in the signals given to the transport user when choosing between transport modes. That includes clarifying and resolving the problem of the different components in the relative prices of rail versus the car or the lorry at the point of use. For a car journey, the motorist tends to compare only the marginal cost (basically petrol) against the price of a rail journey that reflects its total costs, including infrastructure, less any support grants. Thus, even though the average price per mile of rail travel is around 12 pence per mile, against a total cost of a car journey for a single occupant of perhaps three times that, this is not how it looks to the traveller. The cost of rail travel can be reduced further but there is no foreseeable prospect of reducing it to a level where it equates to the variable cost of using the car, and the problem is even more intense in the fiercely competitive market for freight.

There are a number of ways of approaching the problem but none are free from difficulties. Nevertheless, the difficulties need to be overcome if competition between the modes is to be effective in leading to a rational use of resources.

Investment

Investment needs to be maintained. If the balance of spending between the different modes is going to meet the requirements of a sustainable system, then one way or another, wider economic and social costs and benefits need to be reflected in the investment decisions across the transport sector. The fact that we are faced with our present problems of pollution and congestion suggests that we have not been managing to do that effectively.

My prejudices will be understood when I say that I believe that, if we did so, it would lead to a higher priority going to rail investment than has been the case. Certainly, that would be consistent with the experience of other European countries whose past and planned investment in the railway system per head of the population is much higher than ours.

But it is certain that, without a sufficient investment flow, the railways will simply not be capable of contributing fully—or ultimately at all—to the country's transport needs. The investment has to be sufficient to keep the existing network up to date, to allow it to take advantage of new opportunities such as Thameslink 2000 and to finance entirely new railway developments such as the Channel Tunnel Rail Link and Crossrail. Whether the investment comes from the public or private sectors is very much a second-order issue compared with the importance of it coming from somewhere.

In general, these wider issues of pricing and investment seem to be more effectively handled at local level—by the PTEs and other local authorities. At that level, the transport problems and opportunities can more readily be seen in the round. The total journey requirements—from start to finish—of different groups in the traffic mix can be identified and provision can be made for the most appropriate combination of transport modes to meet the total needs. The different modes can be used where their strengths will contribute most. That is valuable not just for the opportunities it can bring for the public transport operators but also for the welfare of the communities involved.

Thus I would see, as a key part of the framework of policies we need, the effective devolution of decision-making on local transport needs to local bodies, together with the necessary funding. In the reorganizations of local government and of the railways I believe it is vitally important that we do not lose the potential for the creative and productive focus on transport that those in the locality can bring. If anything, I should like to see such devolution expanded and enhanced.

At the national level, successive governments have confirmed that they are willing to apply cost–benefit analysis in appropriate cases. Possibly the outstanding railway example is that of the Channel Tunnel Rail Link. To enable that to go forward as an investment by the private sector, the government will be making a free transfer to the successful bidder of the former British Rail assets created by its £1 billion investment in European Passenger Services. Thus the government are demonstrating very clearly both their recognition that there can be very large benefits from such a venture that cannot be captured through the fare box and their acknowledgement that substantial pump-priming can be needed to attract private sector capital into major infrastructure projects.

Cost–benefit framework

The government's approach is a very welcome one. But individual projects need to be part of a cohesive whole if the sum of the parts is to give us the maximum benefit. We know that attempting to provide a blueprint for travel by different transport modes simply does not work. And yet, in our new policy framework, we need to find the mechanisms that will broadly enable each mode to be used to its best—and most sustainable—advantage in meeting people's transport needs. And that means what is best for the different component parts of the journey whether it be for the passenger or for the tonne of freight.

There is no simple answer to the problem. The 1990 White Paper *This common inheritance: Britain's environmental strategy* (Secretary of State for the Environment *et al.* 1990) made it clear that its environmental objectives would only be met if all Whitehall departments considered the environment in the progress of their own work. The problem of transport and the environment is a subset of that. Targets, as suggested by the Royal Commission on Environmental Pollution (1994), would undoubtedly help to focus policy and practical efforts. But they would need to be accepted as well founded ones by the community at large. And, indeed, extending the White Paper's idea beyond the confines of Whitehall, there is the powerful impact of social attitudes more generally.

In our lifetimes, we have seen dramatic shifts in the social acceptability of such diverse activities as drinking-and-driving, wearing fur, and smoking in the presence of non-smokers. Many such shifts are led by government action. But it is social attitudes that do more than anything else to turn policy aspirations into actual changes in personal behaviour. Continued campaigns of information and education could alert people increasingly to the different impact that different journeys can have. Carried out successfully, they could mean that policy measures work with a tide of forward-looking social preference for transport choices that help protect our future. The alternative is for them to work against a wall of narrow-visioned defence of choices that threaten it.

In the course of this chapter I have mentioned many policy elements that have to cohere to produce the outcome the nation needs. However, if there is one step that could yield us major dividends, it is perhaps a radical reappraisal of how the relative merits of different forms of transport are calculated—whether it be for local or for national projects. For some reason or other, the methods we have used failed to warn us that we were going in the wrong direction. They failed to prevent us facing a crisis over pollution, and they failed to tell us that building more roads would not solve the problems of congestion.

One of the reasons may well be that cost–benefit analysis tends to be invoked when we want it to help us prove the case for doing something; it is not generally employed in such a way that it can tell us when not to do something. It may be that our techniques are wrong, it may be that we have not used them as we should, or it may be that we have not wanted to hear what they have been telling us. But, for whatever reason, we do seem to have fallen between two stools. Pure market principles have been recognized as inadequate for giving us an acceptable transport policy—economically, politically, or socially—while market signals have not been modified in such a way as to avoid the environmental and economic crises we now face.

The watershed in policy that we are facing gives us a new, albeit late, chance to change to a sounder course. Many people have a part to play in that process—politicians, transport operators, such as the one I represent, and, not least, policy researchers and thinkers such as are represented at Linacre College. Together we need to get things right both to improve the quality of our own lives and to ensure that we do not impose impoverished ones on generations to come.

CONCLUSION

In conclusion, I hope I have given some idea of the substantial contribution that I believe the railways can make to sustainable transport in Britain. Railways have inherent characteristics that, in the right markets, make them somewhat kinder to the environment than roads. Particularly since the Second World War, the industry has built on those characteristics and made it possible for an essentially Victorian infrastructure to provide an effective base for a modern railway system in the next century. Resources have been used with increasing efficiency. The industry has spare capacity that could be used with little extra cost and negligible additional use of resources to produce significant external benefit.

The policy challenge could hardly be more daunting as people perceive threats to their freedom of action. Nevertheless, changing social attitudes could be an important key to success.

There are fundamental issues still to be addressed in developing a new policy framework for sustainable transport development, including investment, devolved decision-making, and the treatment of social costs and benefits. Even with a favourable policy environment, all in the industry will need to do a great deal of work on many fronts to win business for rail against continuing competition from other modes. It is only by winning customers—rather than by having them thrust upon it—that the industry will maintain the confidence and vitality it will need in order to play its full part in a nation that both promotes economic efficiency and also truly cares for the environment and for the health, safety, and welfare of its citizens.

The railway picture of the twenty-first century may not have guns and dogs or even people in hats. But it should regain an aura of satisfaction and anticipation of a pleasant experience. If that picture becomes a reality, then a real contribution will have been made.

REFERENCES

Department of the Environment/Department of Transport (1994). *Planning policy guidance: transport*, note 13 (PPG 13). HMSO, London.

Royal Commission on Environmental Pollution (1994). *Transport and the environment*, 18th report of the Royal Commission on Environmental Pollution. HMSO, London.

Secretary of State for the Environment, Secretary of State for Foreign and Commonwealth Affairs, Chancellor of the Exchequer, President of the Board of Trade, Secretary of State for Transport, Secretary of State for Defence, *et al.* (1990). *This common inheritance: Britain's environmental strategy*. HMSO, London.

6

The Channel Tunnel, fixed link and positive choice

Alastair Morton

Sir Alastair Morton was educated at Witwatersrand University, Johannesburg, South Africa and took an MA degree as a member of Worcester College, Oxford; he is now an Honorary Fellow of the College. After holding a succession of senior positions in finance and industry, Sir Alastair became the first Managing Director of the British National Oil Corporation (BNOC) in 1976. In 1982, he was appointed Chief Executive and subsequently Chairman of the Guinness Peat Group. Five years later, Sir Alastair began his involvement in the project with which his name will always be associated, the Channel Tunnel: he has been the Co-Chairman of Eurotunnel plc since 1987 and was also Group Chief Executive of the company from 1990 to 1994. If, against current expectation, the Channel Tunnel achieves financial success to complement its undoubted status as an engineering triumph, this will be very much Sir Alastair's personal achievement. He has always taken an active interest in technical and vocational training: he has been Chairman of the Kent Training and Enterprise Council since 1990.

INTRODUCTION

On Monday, 31 October 1994 I was in my office in Calais, on the huge Eurotunnel terminal. Outside a full gale was reaching its climax—perhaps force 9—and it was raining hard enough to suit the Amazon: a tropical downpour carried horizontally across the bleak landscape by the wind. Three miles north of me, tugs were out (as at Dover) assisting those high-sided ferries to struggle in and out of port, hopelessly behind schedule. That well-known symbol of speed across the channel—the hovercraft—had given up completely and gone to bed, abandoning its customers. A few weeks before the continent would have been virtually cut off from Britain—to quote the classic headline of some decades ago. But, half a mile to the south of me, as on every other day of the 2 preceding weeks, between 110 and 120 trains and shuttle trains were passing into and out of the Channel

Tunnel during the 24 hours of Monday. That is less than Eurotunnel will eventually carry and they were not all precisely on time, because we were still debugging the system and the rolling stock during final commissioning, but few if any were significantly delayed by weather, and, in the language of Eurotunnel, 'significantly' means more than 10 minutes' delay—something the airlines at Heathrow do not even allow to be called a delay.

The Channel Tunnel was open and almost fully in business. Even in bad weather it was linking the rail and road systems of Great Britain and Continental Europe 24 hours a day, 7 days a week. A great adventure was approaching completion, 192 years after the first engineering proposal for it was put before a relevant head of state—Napoleon.

By 3 May 1995, the late deliveries of rolling stock and the build-up of demand had almost *tripled* that number of trains and shuttles through the Tunnel every 24 hours, 7 days a week.

Those sleek Eurostar trains, each replacing two of the Airbuses at present cluttering the runways and skies between London and Paris and Brussels, pass from portal to portal in 22 minutes or so, though records have already been established—18.4 minutes portal to portal across the Channel, and a complete journey from London to Paris, city centre to city centre in one seat, in 2 hours and 49 minutes. Our very large shuttles spend 27 smooth minutes in the Tunnel; a favourite trick has been to stand a pound coin on edge during the journey—it stays there all the way.

The fragility of our ferry links has been cruelly exposed in the recent and further past. We live by trade and in the second half of this century our trade has undergone dramatic change. Forty years ago two-thirds of our trade was over the deep oceans from London, Southampton, Liverpool, and so on; today two-thirds takes place by ro-ro ferry and container ship over the short sea crossings to mainland Europe. And the trend will continue. At last, 6 years before the start of a highly competitive twenty-first century, a high-capacity fixed link is in place to carry that lifegiving trade by rail as well as road and to permit us to travel conveniently about our business and our leisure. The Channel Tunnel is open.

A major piece of transport infrastructure has been put into place in the overcrowded yet beautiful south-east of England. Its construction created some disturbance to the environment, though much less than was feared by the ever-lively imaginations of the lobbies, and its operation will shift traffic flows to new channels, road and rail, causing different pressures from those that were there (or not there) before. Truly, the Channel Tunnel is a unique case study in transport and the environment, and I am grateful to Linacre College for inviting me to present aspects of it.

THE FABRIC OF THE EUROTUNNEL

Let me explain a bit about what the Eurotunnel comprises.

1. There are two terminals, one in the Pas de Calais and one in Kent, for transferring road traffic to our shuttles and back on to the road. The French terminal is three times the size of the British, because land was available, and a commercial park and later an industrial park are being built next to our activity in the ample space there. A major national port has been shoe-horned into Folkestone's backyard, disturbing a mere 800 acres of sub-urban Kent.

2. The Tunnel that lies between the terminals, actually three tunnels, was bored through virtually impermeable clay around 100 feet below the seabed—a system 50 km from portal to portal and thus 150 km of tunnel in all. The tunnels are immensely strong and offer safe refuges every 400 yards. They were bored not from each end, but from two coastal sites 38 km apart, with tunnel-boring machines (TBMs) going back to each terminal inland as well as out to sea to meet each other. It was keyhole surgery: everything and everyone to service these 1000 foot long mon-sters—the TBMs and the work-trains hitched behind them—had to go in by train from each end and all the spoil excavated and other waste had to come out the same way. After a poor beginning, it was a magnificent logistical achievement.

3. Each terminal has had to have road and rail connections to the national systems, and all the materials and wastes had to come from or go to somewhere. Those connections involved logistics and represent a con-summate achievement in minimizing the impact of construction on the en-vironment and on the community.

All this was achieved with about as much construction disturbance as that for a major motorway intersection plus a mile or two of link-road. Euro-tunnel, the contractors, and the local authorities managed to work together to avoid shattering east Kent.

THE INITIATION OF THE EUROTUNNEL

I am proud to have played a part in that, so let me say a little about the process. There had been several false starts, most notably in 1880 and 1970, from both of which the British government retreated in disarray—to the disgust of the French.

However, once Mrs Thatcher decided in 1984 to attempt it again, President Mitterrand was willing to trust her determination and join in. Thus, this great project has been initiated and finished in 10 years, with construction and fitting out taking 8 years.

The first steps were to negotiate a treaty with France, then to create a concession structure and framework for considering competing bids by rival promoters, with prompt selection of a winner. Meanwhile, sponsors had to seek parliamentary approval in both countries—in this country via a hybrid bill. All that and the underwriting of a worldwide syndicate loan took until July 1987. This was a hectic and fraught period for all—much more so in Britain than in France, where (for example) the legislation passed each house unopposed, and the high-speed rail route from Paris to Calais was defined in a few months *without* public inquiry but *with* ample compensation.

In Britain, in the winter of 1984–85, a dedicated band of executives, seconded from five British groups of contractors and two major banks, plus attendant consultants, chaired by a retired diplomat (Sir Nicholas Henderson) and in constant consultation with the French half of their bidding group, were hard at work on environmental and transport policy issues as part of their campaign to be the selected concessionaire.

Many of those early pioneers have moved on, or retired, but four were still with Eurotunnel at the end of the project—Tony Gueterbock (now Lord Berkeley), Martin Hemingway, John Taberner, and John Noulton, now with us but then the civil servant at the Department of Transport who had closed and locked the 1970s project and now as Under Secretary was in charge of launching the 1980s project—the treaty with France, the concession to the selection promoter, the legislation through Parliament, and the binational regulatory framework to govern its construction and operation. I joined the group in February 1987.

THE 'ENVIRONMENTAL IMPACT STUDY' AND THE CHANNEL TUNNEL IMPACT STUDY GROUP

In what follows I focus primarily on work led by Tony Berkeley.

He and his colleagues first prepared an environmental impact study in Kent. This was not then a prevalent practice, but it was an essential tool for promoting Eurotunnel's cause. Kent was then the heart of NIMBY ('not in my back yard') land—despite the fact that as a region of England it had prospered for 800 years from its location between the metropolis of London and the rest of Europe.[1]

The environmental impact study nourished a consultative body established by John Noulton and others from the Department of Transport called the Channel Tunnel Impact Study Group, which brought district councils, the Kent County Council, the company, the railways, the relevant Whitehall ministries, and other key parties together. That body was a pioneering and valuable forum, and was only dissolved in 1994. Consultants and internal researchers collected data and examined potential consequences from end to end of Kent, and set an agenda for co-operation, consultation, and supervision that continued throughout the project.

There followed the passage of the hybrid bill put before Parliament in 1986. The process of passing such a bill involves not only intense committee stages in each House of Parliament but also somewhat repetitive hearings by a Select Committee of each house. It is democracy at its best and it is very hard work for all concerned—members, petitioners, experts, and promoters. Amazingly, it took only 9 months to get through the material elements of the parliamentary process, but the public voice was heard and debated. Eurotunnel, as promoter, gave undertakings during the process to respond to environmental concerns that must have cost us well over £50 million.

CO-OPERATION BETWEEN EUROTUNNEL AND THE LOCAL COMMUNITIES

The presentation of an environmental impact study and the preparation of a report by the Channel Tunnel Impact Study Group, and then of the Channel Tunnel Act, did not end the task of Tony Berkeley and his colleagues. Indeed their task is still not finished.

[1] I was able to defuse the Kent problem by pointing out that there were (and are) two Kents, with the line of the M20 from the M25 to the coast the appropriate frontier. South and west of the M20 is leafy, semi-suburban, executive commuterland; comfortable, affluent, and articulate—and remote from our worksites. North and east of the M20 are some stretches of beautiful agricultural Kent, but overwhelmingly in people terms it was a depressed, underskilled region of 'sunset' industrial and commercial activities. When I arrived early in 1987, Thanet, the north-east tip of Kent embracing Margate, Ramsgate, and Broadstairs, had 22 per cent unemployment—matching Merseyside, with the difference that Kent's articulate folk, or 'chattering classes' as some call them, turned their backs and did not talk about it. I stopped that refusal to address the problems. The potential and the economic prosperity of north and east Kent has been on the agenda ever since. From spring 1987, we in Eurotunnel endeavoured only to listen to the worries of Kent on and to the north and east of the M20 frontier. It simplified our task—we had more interests in common with an economic region needing regeneration, even if the Canterbury District Council and some members of parliament could not see it.

For example, there were three small villages adjacent to our Folkestone terminal: Frogholt, Newington, and Peene—quite apart from the outer Folkestone suburb on the other side of the M20. Tony Berkeley and his colleagues addressed the issue of these villages even before Eurotunnel was selected as the concessionaire in preference to the other bidders. The scheme they devised, in consultation with local people, was an outstandingly successful model that the public sector has stubbornly refused to follow in the succeeding 10 years. Its cost and cash-flow effect on our finances was much less than the ungenerous purchase and compensation practices permitted by the Treasury.

Tony Berkeley, John Taberner, and their advisers held public meetings in each village and at the end of the process had created a system whereby local residents had only to bring in the title deeds to their home at any time in the succeeding 10 years—that is, until about now—to receive in cash, without argument, a sum known to them, or calculable by them, from 1984 onwards. The sum started at a generous premium over unblighted market value, that is market value if there were no tunnel scheme, and moved thereafter according to a known index. Its simple concept permitted the residents to wait and see how bad the construction or the operation of the Tunnel proved to be for their lives. They did not *have* to sell, then or ever; they did not *have* to sell into a blighted market or at blighted prices. In the event, by the time construction peaked in the early 1990s, something like 50 per cent of the villagers had sold to us, without disputes. The flow of sales pretty well stopped before that when the villagers discovered that Eurotunnel was able to sell the houses on the open market at prices above the generous formula price, given the property boom of the time and thanks to Eurotunnel's success in limiting the noise and nuisance from the project—not least by a large embankment. Thus our cash flow out was deferred and reduced by 50 per cent: how simple, and how fair to the local residents—and, therefore, naturally, how difficult for the Treasury to understand.

Moving on from there, Eurotunnel co-operated with John Noulton and his colleagues in Whitehall in the establishment of Sir Donald Murray as Channel Tunnel Complaints Commissioner, based in Folkestone. His task was to be available to any local residents or concerns; to listen to any problems concerning the way in which the project was being carried out; and then to ensure that the problem be taken into consideration and, if possible or necessary, action be taken by contractors; Eurotunnel, local authorities, or whoever was responsible to reduce it. Donald Murray is a former diplomat, not without humour. To save public money, he wisely established his office in the detention cells next to the Folkestone magistrate's court; people of bad conscience tended not to make a nuisance of themselves there, but he provided the essential safety valve just by being there, however bored he complained later he had become. The most usual complaints were noise, mud on the road, fenceposts knocked over, and so on. They were heard and dealt with, if necessary, by village hall meetings attended by both Eurotunnel and the contractors.

The Folkestone terminal is wholly within Shepway district, comprising the towns of Folkestone and Hythe, plus Romney Marsh, whereas on the French side the French terminal is intruded upon by five communes each with its own mayor. And a French mayor is a political person with clout, even when his commune is a tiny village in the shadow of a large town like Calais. Eurotunnel's relations with the Shepway District Council were at all times close, even if sometimes combative, but not confrontational. I regarded it as part of my job to ensure that the political leader of the District Council talked regularly with me while his or her officials battled out the issues with my people. During the project the leadership went from Tory to Liberal Democrat, but the co-operation did not change. We also had to co-operate with Dover District Council, particularly over coastal and tidal matters. Dover and Dover Harbour Board, the legal guardians of coastal or tidal matters in that neighbourhood, stood to be severely injured over years to come by our competition with the ferries. We found the occasional official who thought he had to use his public office to fight the ferries' battle for them, but, in general, we benefited from the high standards of local public servants in the discharge of their official responsibilities.

When I welcomed the Queen back to Folkestone on 6 May 1994 as the first British sovereign ever to return from a foreign adventure by land, I went on to pay a tribute to such people in the following terms:

> Let me also, here in Kent, pay tribute to another layer of partnerships—between on the one side the private sector interests taking this end of the project forward year after year, as contractors or as owners, and on the other side the elected leaders and officials of government, in Shepway, in Dover, in Kent and nationally, who have worked tirelessly to achieve—with fairness and consideration to all—the very large change in local, and even national, circumstances that is the Channel Tunnel.

These relationships and partnerships have to be built, but I do not regard it as unfair that enterprises such as the Channel Tunnel should be required to build them. It is the interface of commerce and democracy, and we need to improve our performance at that interface if we are to maintain our environment while we meet our transport and other modern social needs.

As an extension of their work with local residents and other organizational and 'people' concerns, Tony Berkeley and his colleagues went on to encourage the bedding down or taking root of the Channel Tunnel in the local community—that is, of the future transport service operated by the private sector Eurotunnel. After the construction workers left, we were to be a transport operator employing over 1000 in east Kent, which is not a region of large companies. Meanwhile, and in preparation for the adjustment, whether mental or physical, required for miles around, we had to avoid not just irritation but the beginnings of any potential injury. There were changes other than simply the increase of traffic on the roads, which we had made every effort to minimize during construction.

We established a successful exhibition centre alongside the motorway next to the Folkestone terminal. Throughout the 7 years prior to opening 250–300 thousand visitors a year came to learn about the Tunnel, its environment, and its future operation—over 60 per cent were school groups.

Next, Eurotunnel people developed partners, liaison groups, and simple contacts with the village and town, and with the academic and other communities of east Kent. Many villages now have minor improvements to their village halls to show; many colleges have had programmes supported; many others have memories of supported community events. The artistic life of east Kent received a friendly push on a number of fronts and indeed 'Primavera', an east Kent chamber orchestra that we rescued from financial oblivion in 1987, is now well known across Europe.

The 'Célébration '94', community events in east Kent in May 1994, most of them involving local community and artistic activities attended by tens of thousands of visitors and locals over a period of weeks, were the entertainment and cultural peak of years of friendly co-operation. Ah, 'bribing them with sweeties', I hear some say. 'Not true', I reply. It's business in (and for) the community, lifting people's eyes and ears to the way things are changing.

When I arrived in spring 1987, I looked back to my experience in the mining industry in central Africa when, as the biggest and most forward-looking local employer we worked to develop a medium- and long-term vision of the future economic structure to be made possible by my company's activities. As a result of that experience, I said in 1987 that, whatever the outcome in detail, east Kent would not be the same after the Tunnel project. It would be more open to people from across the Channel, it would be more prosperous, it would need the people and the social and physical infrastructure to make the most of its new advantages, and these would be the *quid pro quo* for giving up the rural isolation, increasingly poverty stricken, that they knew when they were simply the backyard of that decaying imperial metropolis, London.

In addition to this community relations programme, jointly with the County Council's Education Department, we also created an outreach programme into the schools of Kent and beyond. For 8 years, school project study programmes and related language and other educational packages have been jointly developed by Eurotunnel, the BBC, the County Council, and others.

When the Government founded Training and Enterprise Councils (TEC) to put skills development, training, and the development of small business under the guidance of private sector-led councils, I helped to found the Kent TEC and have been its Chairman since. When it became desirable to get Assisted Area Status for the depressed parts of east Kent, Eurotunnel and the County and District Councils brought together the East Kent Initiative, private sector-led, with its first Chief Executive seconded from Eurotunnel and with me as non-executive Chairman. We succeeded in our aims.

We have also published some of the story of the archaeological and the countryside management programmes that have safeguarded and restored rural scenes in our neighbourhood.

There was more, but what I have tried to show is that the social environment rated as high in Eurotunnel's transport project as the landscape did; and neither was resented or pushed aside as unimportant by a company that had to be created from scratch for the project and that was built up under the maximum of pressure from many conflicting interests, not all financial.

CONCLUSIONS

What conclusions can be drawn about transport and the environment as seen through Eurotunnel's telescope? In summary, these are that the private sector can handle these issues, should be trusted by communities, governments, and shareholders to handle them, and that their considerable cost to private enterprise is a legitimate cost of the project provided it remains in proportion and is not driven through the roof by delay or malevolence such as can come from the efforts of hostile parties—some of whom have legitimate claims to an interest, though many do not.

I am sure there are limits to the optimistic picture I am painting. I believe that those limits are much wider than some investment interests or government officers would believe. The difference at Eurotunnel was the initiative and effort of some dedicated people for whom I had, and have, the honour to speak. The sanction for their efforts had to come from the top. Eurotunnel has invested in the environment surrounding this project—in the people and in the social and the physical environment. I see nothing to complain of when that is asked of a megaproject.

And now, since the recent Royal Commission on Environmental Pollution report (1994) has come out to stimulate our thought processes, let us seek to bring forward the investment projects that will lighten the burden of transport pollution on our environment. In other places I have defined that pollution as taking four forms—emissions, noise, casualties, and land-take. I am pleased to see it proposed that we think this through seriously. I hope we will.

And so let me end by paying brief lip service to another aspect of the title for this chapter, which I selected months ago, 'The Tunnel, a fixed link and positive choice'. I have talked about the positive choices we took about our megaproject, the fixed link, and how those who come after can take a similarly positive line with their projects.

Also important are the positive choices we can and *should* make about how we transport ourselves—road versus rail, rail versus air, water versus pipe versus rail and tunnel or bridge. Just why did the Swiss people vote in a referendum, in a ratio of 5 votes to 3, to build a 50-km tunnel to keep freight on rail and off road 15 years from now? The choices are there: let's consider them.

REFERENCE

Royal Commission on Environmental Pollution (1994). *Transport and the environment*, 18th report of the Royal Commission on Environmental Pollution. HMSO, London.

7
Airlines, aviation, and the environment

Hugh Somerville

Dr Hugh Somerville took his first degree, in chemistry, at the University of Edinburgh and a PhD in microbiology at the University of Sheffield. He undertook post-doctoral research at the University of California, Los Angeles, before spending a year as an Assistant Professor at the University of California, Berkeley. Dr Somerville then joined Shell Research and was appointed Head of the Microbiology Research Division. He has held environmental posts with Shell in Houston, The Hague, and Aberdeen as well as with Occidental Petroleum. In 1989, he joined British Airways to take up the newly created post of Head of Environment, which he still holds. He has been active in a number of environmental groups and is currently (August 1995) Chair of the International Air Transport Association (IATA) Environment Task Force.

INTRODUCTION

This chapter reviews the relationship between the airline industry and the environment in general terms rather than with reference to the specific policies of British Airways. It is noted that other chapters in this volume include road transport, the railway industry, the Channel Tunnel, integrated urban transport systems, and marine transport. In the present contribution, the environmental issues facing airlines and the aviation industry will be identified, along with the ways in which they are being addressed within an overall regulatory framework. Some selected references to other sources are made, to allow more detailed examination of the issues.

Worldwide, a fleet of over 10 000 civil aviation aircraft flies more than a million million passenger kilometres each year. There are various predictions of growth, mostly around 5–6 per cent for the foreseeable future. As one part of this industry, British Airways operated a fleet of over 240 aircraft with 291 000 flights in 1993–94. There are 50 000 employees and, with the global alliances with other airlines, some 95 million passengers are flown each year to 445 destinations in 85 countries.

While British Airways is the world's largest international airline, it is still small compared to other carriers—being fifth in terms of passenger kilometres and ninth in terms of total passengers carried (IATA 1993). Likewise, in terms of passengers handled, Heathrow, Gatwick, and Manchester airports rank fifth, twenty-eighth, and forty-fifth, respectively, on a worldwide scale.

It is self-evident that an industry of this size has some impact on the environment and I shall proceed directly to review the environmental issues, which may be conveniently discussed under the following headings:

- noise;
- fuel efficiency and emissions to the atmosphere;
- congestion and infrastructure;
- waste of energy, materials, and water;
- tourism and conservation.

These are the issues identified through an environmental audit of our operations and facilities at Heathrow and the worldwide flying operation (British Airways 1991). This has been confirmed by research through in-depth interviews with a number of opinion leaders in different sectors. When the interviewees were asked unprompted to identify the major issue, 'noise' was an almost automatic response whereas, when presented with a list, emissions to the atmosphere emerged as most important.

These results indicate that different groups have varying perceptions of the relative importance of environmental issues. For example, when asked to distribute 100 points of 'relative importance', tourism pressure groups attributed 48 points to tourism and only 14 to noise, whereas noise pressure groups attributed 47 to noise and 4 to tourism (Table 7.1). Perceptions are important in the environment, partly because of the difficulty of reducing different environmental impacts to a single currency, but also because different individuals and groups hold differing views and such perceptions have to be treated as real until the perception is changed.

NOISE

Overall impact of aircraft noise

Aircraft noise is recognized as a major nuisance to many people living around airports. The industry has responded by introducing measures to

Table 7.1 Relative importance attached to environmental issues by several interested groups*

Issue	Relative importance (%) of issue to groups interested in (or comprising)					
	Noise	Tourism	Academics	Chambers of Commerce	General environment	Customers
Emissions	17	10	41	20	24	25
Noise	47	14	13	18	19	24
Waste of energy, etc.	14	16	29	15	24	29
Congestion	18	12	7	36	12	14
Tourism	4	48	11	11	21	10
Total	100	100	100	100	100	100

* Source: British Airways.

Table 7.2 Trends in community noise exposure at Heathrow Airport

Year	Population* within 57 LEq (35 NNI) noise contour†
1974	2 004 000
1976	1 977 000
1978	1 467 000
1980	944 000
1982	1 028 000
1984	776 000
1986	695 000
1988	538 000
1990	488 000
1991	429 000

* The population numbers are based on the latest current census. For example, for 1991 the base changed from the 1981 census to the 1991 census.

† The basis for the contour changed from the noise and number index (NNI) to LEq in 1989 with only a marginal effect on population numbers. Both are measurements of overall integrated noise from aircraft movements. 57 LEq (35 NNI) is defined as the 'onset of disturbance' by the UK Department of Transport. LEq is the equivalent continuous sound level covering a 16 hour day, that is excluding noise from night-time movements of aircraft.

control noise at source which have resulted in a substantial reduction of the noise impact both of single aircraft and of the integrated noise arising from overall aircraft movements (Smith 1989). There has been an enormous decrease in the 'footprint', that is, the area receiving a given noise level from a single aircraft movement, for example, when newer models of a single aircraft type are compared with their earlier counterparts. This translates into a reduction in the overall impact of noise at most airports.

For example, at London's Heathrow airport the number of people living within the area receiving the noise level of 57 LEq has fallen to about 20 per cent of that receiving the same noise 20 years ago (Table 7.2; CAA 1994). While such integrated measurements of noise do not indicate the noise of individual movements, they do reflect the enormous investment of the airlines in newer quieter aircraft. A recent study by the manufacturing industry (ICCAIA 1994) found 'an identifiable relationship between noise and cost to society, both in terms of the price paid to transport people and goods by air, and in the production of engine exhaust emissions through the amount of additional fuel burned'.

Night noise

Noise has many different meanings, and it is not possible to describe its impact totally in scientific terms. This is particularly true of noise at night.

Recently, the Department of Transport has sponsored a study into the effects of aircraft noise on sleep (Department of Transport 1992). The conclusions of the study included the comment that 'once asleep, very few people living near airports are at risk of any substantial sleep disturbance due to aircraft noise even at the highest noise levels'. Nonetheless, night movements are a source of nuisance to many who live close to airports. Many smaller airports operate a night curfew and most large international airports restrict night movements.

Unfortunately, movements at night are a requirement for an effective international aviation network. While rules and regulations vary from country to country, overall there is significant sharing of this impact and airlines make every effort to avoid unnecessary night movements.

Prospects for noise

The improvements over the last two decades have largely resulted from the introduction of high bypass turbofan engines. In bypass engines part of the air passing through the engine does not enter the combustion chamber and the combination of the bypass air with the exhaust from combustion leads to a less turbulent stream, thus reducing noise. This rapid improvement will carry on for a few years but will not be indefinite. The phase-out of older noisier 'Chapter 2'[1] aircraft is nearing completion, against a regulatory deadline of 2002 (which allows for a few exceptions). The opportunities for further improvement in aircraft noise are reaching the stage of diminishing returns, whether with respect to noise from engines or airframes.

The main contribution to decreased noise will come from phasing out of these Chapter 2 aircraft, such as the British Aircraft Corporation BAC 1-11s. Twin-engined aircraft also help to reduce noise impact, as they climb out more rapidly. Hush kits, retrospectively fitted to aircraft engines to reduce noise as a means of upgrading Chapter 2 aircraft to Chapter 3, are of debatable environmental benefit. Inevitably, they increase the weight leading to decreased fuel efficiency and an increase in emissions. This is, however, an option that will be used to keep aircraft in service. Re-engining is also a possibility and there is still scope for noise reduction through improved technology. These improvements have to be weighed against increases in aircraft size and any increase in the number of movements.

No presentation on noise is complete without mentioning the need for greater attention to use of land around airports. An illustration is Hounslow

[1] 'Chapter 2' and 'Chapter 3' refer to noise certification standards set by the International Civil Aviation Organization; see the section of this chapter on 'Regulatory background' for further detail.

Heath over which a departure route from Heathrow was planned as a minimum noise path, only to be followed by planning permission and the building of a large number of new houses immediately under that flight path. Recently, UK government guidelines have been strengthened (Department of the Environment 1994*a*) but there is still doubt over the effectiveness of planning controls on residential development close to airports.

WASTE

Waste is an issue for any organization. As this is a general issue it is not discussed here in detail. The aviation industry has some particular problems, for example, in disposal of in-flight waste, de-icing, and in aircraft painting. As in other sectors, environmental interests clearly coincide with financial interests. Savings in waste almost always result in a positive effect on the bottom line. For more detailed information on waste see British Airways (1994).

CONGESTION

Congestion is an environmental evil in that the consequences of inadequate infrastructure, whether in the air or on the ground, are inevitably deleterious to the environment. From the viewpoint of aviation this can be divided into two areas.

Congestion on the ground

There is growing recognition in the aviation industry that it has a significant part to play in minimizing congestion on the ground. It is important to customers that they should enjoy efficient access to air travel. Heathrow Airport Limited already estimates that over 30 per cent of its passengers travel by public transport (Table 7.3; Heathrow Airport Limited 1994). A tough target of around 50 per cent has been set for the future, albeit against a background of overall growth. The success of Metrolink in Manchester is another example. At Heathrow, the airport is investing, through a joint partnership with British Rail, in the Heathrow Express due to start service in December 1997. The airport has also invested heavily in providing facilities for bus access, to the point where the bus station in the central area is arguably the busiest in Britain and many bus travellers make bus connections without taking a flight. All of the indications are that the time is now

Table 7.3 Public transport access to airports*

Airport	Passengers/year	Passengers (%) travelling by: Public transport — Train	Coach /bus	Total	Other modes of transport†
UK airports					
Heathrow	47 600 000	20.2	14.9	35.1	3.5
Gatwick	20 100 000	24	12	36	–
Birmingham	3 700 000	5.5	8.4	13.9	5.2
Manchester	11 700 000	3.5	12.4	13.9	2
Airports in other countries					
Schiphol (Amsterdam)	20 800 000	27	3	30	3
Charles de Gaulle (Paris)	25 700 000	15	18	33	–
John F. Kennedy (New York)	26 100 000	–	7.7	7.7	7.7
Narita (Japan)	20 000 000	24.3	24.3	48.6	5.8

*Reproduced, with permission, from a more detailed survey by Heathrow Airport Limited.
† 'Other modes' include cycling and walking.

right to move towards more effective use of ground transport serving airports. This feeling has been strengthened by the recent report from the Royal Commission on Environmental Pollution (1994).

Ground congestion problems also include delays in departure and in the period between landing and shutdown of engines. Inadequacies in infrastructure such as limitations in terminal capacity at Heathrow contribute to such delays.

Congestion in the air

Almost everyone who has flown commercially in the UK will have experienced 'stacking'. These holding delays, which reflect the unavailability of landing slots for one reason or another, cost British Airways some 50 000 tonnes of fuel per year at Heathrow and Gatwick alone (British Airways 1994). While much fuel is burned in the stack, a significant part is burned as a result of carrying the extra weight of the fuel to provide for the eventuality of being stacked. This arises from delays in air traffic control (ATC) and air traffic movements (ATM). Other airlines also suffer. Overall a recent study for ATAG (the Air Transport Action Group) has indicated that the cost of congestion on European airports and airspace would be about $6 billion by the year 2000 (Meredith, personal communication 1994; for information contact ATAG, Geneva, Switzerland). The International Air Transport Association (IATA 1991, p. 21) reported that 'Europe has 51 air traffic centres using 31 different ATC systems, 22 different management systems and 33 different computer languages'.

Factors producing delays include inefficient flight routings, military airspace constrictions, and low ATC and ATM system productivity. Since 1990 a number of improvements have been initiated including actual or planned measures at Heathrow to increase runway and terminal capacity. Through the ECAC (European Civil Aviation Conference) improvements in Europe have included the evolution of the 'Central Flow Management Unit', which should be fully operational in 1995, offering pilots optimum routes and altitude levels to achieve the completion of their flights with the least possible waste in time and fuel. Airlines, including British Airways and Qantas, have started flying and testing packages designed for the Future Air Navigation System (FANS), which will rely on satellite-based technology to provide a global and seamless system. In February 1994 the importance of action in this area was reinforced when the European Commission's 'Comité des Sages' called for an end to the fragmentation of ATC systems in Europe.

EMISSIONS AND FUEL EFFICIENCY

Fuel efficiency

The record of civil aviation in improving the fuel efficiency of transport is outstanding. There are many factors that contribute of which the most important is engine development. The overall fuel efficiency of jet engines has risen from under 10 per cent in the first models to over 35 per cent at present and up to another 10 per cent improvement is projected. Other contributions have come from:

- weather forecasting;
- route selection—often affected by political considerations;
- flight planning systems both in terms of optimizing flying heights and in managing fuel;
- weight minimization and distribution;
- aerodynamics—wing tips, flaps, trailing edges, and engine pods;

Table 7.4 British Airways—trends in fuel consumption and efficiency

Financial year	Fuel use (million tonnes)	Fuel per revenue passenger kilometre relative to 1974/75
1974/75	2.33	1.00
1976/77	2.44	0.95
1978/79	2.79	1.00
1980/81	2.72	0.69
1982/83	2.24	0.62
1984/85	2.39	0.74
1986/87	2.61	0.67
1988/89	3.60	0.78
1990/91	3.78	0.65
1992/93	3.92	0.63
1993/94	4.28	0.62

- laminar flow technology;
- lightweight construction;
- engines with higher bypass ratios;
- combustor technology.

As with noise the historical record of improvements in fuel efficiency is impressive (Table 7.4; British Airways 1994). Over the last 20 years there has been close to a doubling of efficiency in terms of fuel consumed per passenger kilometre. There is confidence in the industry that there will be a continuation in this trend. While there will continue to be a strong financial incentive, it seems probable that environmental considerations will play an increasing role—possibly reflected through economic measures.

Air travel can compete with other modes in terms of efficiency (Table 7.5). Although it is difficult to compare like with like, the aircraft figures shown in Table 7.5 measure actual consumption from flight data systems, compared with figures published by the European Commission (1992) and the Royal Commission on Environmental Pollution (1994). Other factors suggest that the comparison could be more favourable to aviation. These include: the measurement of distance, presumably as actually travelled for road and rail, but direct point-to-point for aviation; the energy and emissions relevant to fuel refining from crude oil, lower for aviation kerosine than for gasoline; the cargo carried by many aircraft in addition to passengers; and the land take, which, for example, in Germany for airports is

Table 7.5 Some relative fuel efficiencies*

Mode of transport	Journey	Occupancy (%)	Fuel efficiency (megajoules per passenger kilometre)	Source†
Train				
British InterCity 225	Unspecified	50	1.0	RCEP 1994
Electric high-speed	London–Paris	50	1.25	EC 1992
High-speed	Brussels–Paris	50	1.43	EC 1992
Automobile				
Small diesel car (1.8 litre)	Unspecified	35	1.2	RCEP 1994
Large petrol car (2.9 litre)	Unspecified	35	2.8	RCEP 1994
Airplane				
Unspecified	Internal flights	65	3.5	RCEP 1994
Boeing 737–400	London to and from Edinburgh	70	1.75	British Airways
Airbus 320	London–Nice	70	1.23	British Airways
Boeing 727	Unspecified	75	1.94	EC 1992

*Source: British Airways.

†RCEP, Royal Commission on Environmental Pollution; EC, European Commission.

Table 7.6 Emissions at Heathrow of hydrocarbons from British Airways* operations

Source of hydrocarbons	Estimated quantity (tonnes per year)	Proportion reaching local environment
Fuel jettison	500	Virtually none
Leaks	8	Small
Spillages	4	Small
Aircraft exhausts	475	Small
Aircraft auxiliary power	3	All
Fuel loading–vapour displacement	10	All
Engine testing	15	All
Heating plant	<1	All
British Airways ground transport	150	All

* Source, British Airways.

less than 8 per cent of that taken for rail and less than 1 per cent of that taken up by roads (IATA 1991).

Emissions

Emissions to the atmosphere from aviation can have an impact at the local, regional, or global level.

Local air quality

On the ground and at the local level there are a number of sources related largely to engineering work and to the large ground fleet of vehicles that airlines operate. For example, British Airways has a fleet of some 3200 vehicles at Heathrow, of which close to 600 are powered by electricity.

The most significant local inputs from aviation activities are from aircraft, the ground vehicle fleet, and from specific engineering activities such as aircraft painting. In an internal assessment (Table 7.6), British Airways has concluded that the ground vehicle fleet is the most significant source of hydrocarbons reaching the local environment.

The general consensus from studies at a number of airports (for example, Raper and Longhurst 1990; Swissair 1992) is that, while there may be small areas that experience significant levels of contamination, air quality at airports is, if anything, better than that in surrounding urban areas. Heathrow Airport Limited have recently reported that all measurements meet EC and WHO air quality standards and guidelines (Heathrow Airport Limited

Table 7.7 Annual emissions of carbon dioxide*

	Tonnes per year (millions)
Worldwide	
Total—fossil fuels	20 000
Civil Aviation	500–600
British Airways	12.7
United Kingdom	
Total	562
Power stations	185
Road transport	106
Civil aviation (up to 1 km altitude)	<1

*Sources: British Airways and Department of Environment (1994*b*).

1994). However, it is realized in the aviation community that there must be a proactive approach in this area.

Fuel jettisoning

An area where there is a common perception of impact on air quality is fuel jettisoning. The jettison of fuel is required occasionally in order to decrease to the safe landing weight. This procedure may be invoked for safety reasons if it is considered necessary to return to the departure airport shortly after take-off. It involves flying to 6000 feet and well away from centres of population. In many modern aircraft the maximum take-off weight and safe landing weight are so close that there is no facility on board to jettison fuel. Studies carried out by flying through dispersing fuel have indicated that only a small fraction of the fuel may reach the ground (Clewell 1983).

Engine emissions—carbon dioxide

An estimate of the overall production of carbon dioxide from aviation can be obtained from estimates of fuel consumption. For aviation, most estimates are derived from fuel consumption and are in the range of 500–600 million tonnes carbon dioxide per annum (for example, Egli 1991).

Carbon dioxide is the principal global warming gas and is thought to account for some 50 per cent of anthropogenic global warming (IPCC 1994). From Table 7.7, it can be deduced that aviation, with less than 3 per cent of emissions from fossil fuels, is responsible for up to 1.5 per cent of global warming from such sources.

Nitrogen oxides (NOx)

For the present purpose no distinction is made between the different species of nitrogen compounds. The very advances in engine technology that have led to greater fuel efficiency have depended on higher temperatures of combustion. Thus, the substantial reductions in emissions of hydrocarbons and carbon monoxide have not been matched by similar reductions in NOx in exhaust gases. The NOx is derived almost entirely from nitrogen and oxygen in the air drawn through the combustor.

Environmental interest in nitrogen oxides from subsonic aircraft has increased in recent years, particularly with respect to emissions at cruise altitudes in the upper troposphere and lower stratosphere (Egli 1990; Schumann 1994). Because of the variation in location of the tropopause it is difficult to be precise on the proportion of exhaust emissions reaching the troposphere and stratosphere. Internal estimates from British Airways, Lufthansa (Reichow, personal communication), and other sources (for example, Schumann 1994) are consistent with, overall, some 30 per cent (20–50 per cent) of aircraft flying time being spent in the stratosphere. For some specific routes such as the North Atlantic, this may be as high as 40 per cent, as a result of the variation in the location of the tropopause and in the flight profiles of aircraft.

The concerns are that, in the troposphere, NOx from aircraft at cruise altitudes, which is emitted above cloud level, leads to the formation of ozone, which can contribute to global warming (Johnson *et al.* 1992). There is concern that aircraft emissions in the stratosphere could lead to depletion of ozone, as a result of differing but also complex chemistry. The ozone generated in the troposphere is neither quantitatively sufficient or in the necessary locations to compensate for stratospheric ozone depletion (Schumann 1994).

Various estimates have been made of the fate and effects of NOx from aircraft at high altitudes and it is an area where much research remains to be done. The most recent authoritative opinion can be found in the latest statement from IPCC (1994, p. 23): 'Our current best guess is that the positive radiative forcing due to the release of NOx from aircraft could be of similar magnitude or smaller than the effect of carbon dioxide released from aircraft'. Thus, along with the impact of carbon dioxide, emissions from aircraft account for less than 3 per cent of man-made global warming effects.

It is, in some opinions, too early to make an estimate of the extent of any corrective action that may be necessary. However, NOx formation is an area where technological fixes may be available. Some are already proven or partly proven, such as the two-stage or double annular combustors fitted

Table 7.8 British Airways fuel consumption and emissions 1993–94*

	Total tonnes per year
Fuel consumed	4 040 000
Carbon dioxide	12 700 000
Water	5 000 000
Hydrocarbons	9 500
Carbon monoxide	3 900
Nitrogen oxides	48 000
Sulfur dioxide	2 400

*Source: British Airways.

to engines that are coming into service with Swissair and British Airways within the next year.

Research into the fate and effects of NOx is being carried out both in Europe and in the USA. The European AERONOX programme (Miles 1994), which is due to report soon, is looking at wake chemistry, distribution, modelling and the overall fate and effects of NOx emitted at cruise altitudes. The study is linked to a database on aircraft routes and quantities of emissions. The MOZAIC programme involves Airbus 340 aircraft of Lufthansa, Air France, and Austrian Airlines carrying measuring equipment for ozone and water vapour along commercial routes. Japan Airlines is also involved in measuring air quality at high altitudes (Japan Airlines 1994).

Other exhaust gases from aircraft

There are other emissions from aircraft. Approximate global quantities of sulfur dioxide, water vapour, carbon monoxide, and hydrocarbons can be estimated from Table 7.8, assuming that British Airways accounts for some 2.5 per cent of the total. The interest in these, for example in the possible effect of water vapour through formation of condensation trails, is less than in the impact of carbon dioxide and NOx. For further information, recent papers can be found in Schumann and Wurzel (1994).

TOURISM

Tourism has been identified as the world's largest industry by the World Travel and Tourism Council (WTTC 1994). WTTC claims that tourism accounts for one in nine jobs on a worldwide basis. Tourism is important to airlines such as British Airways as some 60 per cent of the airline's passenger kilometres are flown by leisure travellers. Indeed every one of the passengers is a tourist by the definition of the World Tourist Organization. The tourism industry, which has many components, is growing rapidly, particularly in the Asia Pacific area. The environmental impact of tourism reflects the breadth of its components, from airlines to ski resorts and from package trips to camping safaris.

One useful analogy is that of the goose and the golden egg, with those employed in the industry feeding on the egg laid by the tourism goose, which is in its turn feeding on the environment. There is a growing recognition that the industry must play its part in nurturing the environment, whether social, cultural, built, or natural. This has been reflected by the formation of the WTTC and the associated Environment Research Centre, in which airlines including British Airways have taken a leading part. Within British Airways, this has been accompanied by a long-standing programme assisting conservation (British Airways Assisting Conservation), a world-wide award scheme recognizing environmentally responsible tourism projects (British Airways Tourism for Tomorrow Awards), a competition for aspiring journalists (Worldwatch Assignment), and inclusion of our tour operator subsidiary in the internal environmental audit programme.

In many ways the impact of the tourism industry is an amalgam of the other issues that we have identified. In particular, the provision of adequate infrastructure is essential to the environmental management of the industry. In many cases the separation of environmental inputs from their impacts has not been made. For example, the same amount of sewage can have very different environmental impacts at two contrasting tourist destinations.

There are difficulties in assessing the overall environmental impact of tourism, largely arising from the difficulty of separating individual components of the industry. For a more detailed discussion, reference can be made to specific publications (for example, Eber 1992).

REGULATORY BACKGROUND

Aviation is subject to a full range of environmental regulations. These may be international, national, or self-regulatory in character.

International regulation

As a result of the Chicago Convention on International Civil Aviation of 1944 the International Civil Aviation Organization was established and, in turn, ICAO established CAEP, the Committee on Aviation Environmental Protection. ICAO first introduced noise certification standards in 1971. Most subsonic jet aircraft now comply with the standards of Chapter 3 of the Annex 16 to the Chicago convention. For example the overall IATA fleet and the British Airways fleet were 61 and 77 per cent Chapter 3, respectively, on 31 March 1994. In November 1993 ICAO ratified a recommendation, supported by airlines, to reduce the limit for emissions of nitrogen oxides by 20 per cent. A number of working groups of CAEP, involving inputs from government and industry, have reviewed the environmental impacts of aviation with a view to possible consideration of increased stringency in engine noise and emissions. CAEP met in December 1995. The outcome is uncertain and has been referred to the ICAO Council which meets in the spring of 1996. Key aspects in the deliberations of these groups are the effects of reductions in noise on emissions as mentioned above, and the fate and effect of emissions at cruise altitudes.

International regulations also emanate from the European Union through the European Commission, or from international agreements covering specific areas, such as the Montreal Protocol and the various subsequent agreements relating to ozone-depleting substances. Most such regulations are concerned with issues that are not aviation-specific, such as water or air quality, and are part of the general regulatory framework. Recently, there has been a move by the EC to penetrate regulatory ground hitherto occupied by ICAO.

National and local regulations

These can be aviation-specific, such as those affecting night movements, preferred routes, some aspects of planning legislation, and other nationally or locally regulated aspects of air transport. Each airport can act as its own local regulator by introduction of procedures that are mandatory for airlines using its facilities. National and local regulations also cover more general aspects such as waste disposal. However, there are aviation-selective aspects to waste disposal; for example, the UK requirements for the disposal of international catering waste.

Self-regulation

It could be said that the early move by airlines to replace Chapter 2 aircraft ahead of the regulatory deadline is a form of self-regulation. In the field of general aviation, some airfields have successfully reached agreement on operational patterns, and at larger airports adherence to preferred routes could also be said to fit this category. Airlines operate noise abatement procedures on take-off and approach to many airports, not all of which are mandatory, and airlines and airports do work together to improve performance for example in noise- and track-monitoring systems installed at a range of airports including Copenhagen, Manchester, and Heathrow.

Economic measures

Often regulation of aircraft noise is accompanied by use of economic measures. There are differential charges, based on noise, at a large number of European airports, and some airports additionally impose fines based on information from the noise- and track-monitoring systems. Overall, these noise charges are a significant cost—for example, British Airways paid more than £3 million in 1993–94. Noise charges are only one of the many charges reflected in the cost of air travel. It is to be anticipated that such economic instruments may be applied more widely in the future. This view is supported by the introduction of emission charges for internal flights in Sweden and of waste regulations, with an option of charges, in Austria.

While no commercial organization welcomes additional charges or taxes, it is recognized that economic measures do have a role to play in the improvement of environmental performance, alongside a realistic regulatory framework and enlightened self-interest.

The Royal Commission on Environmental Pollution

The Royal Commission on Environmental Pollution (1994) considered aviation along with other transport modes in its eighteenth report. Recommendations to the government included:

- to press ICAO to extend emissions regulations to include cruise (all phases of flight);
- to support more stringent noise certification if this can be met without a fuel penalty;
- to collaborate in research into the possible effects of supersonic aircraft on the stratosphere;
- to negotiate internationally for a levy on airline fuel purchases.

These topics are already on the agenda of CAEP and other relevant organizations. Perhaps more controversial was the recommendation to discourage air travel for domestic and near-European journeys. In particular, this recommendation may be based, at least in part, on a false impression of the relative fuel efficiency of aircraft, as discussed above.

While the major concern of the Royal Commission on Environmental Pollution and of many other groups is with transport on the ground, there has been growing interest in the environmental impact of aviation. The specific aspects of emissions from aircraft have been comprehensively reviewed by Archer (1993). The World Wide Fund for Nature (WWF) has addressed tourism (Eber 1992) and aircraft emissions (Barrett 1994) and, in the USA, the Environmental Defense Fund (Vedantham and Oppenheimer, 1994) has also addressed the impact of aircraft emissions. In the UK, the Airfields Environmental Federation (1994) has started to play a role in bringing stakeholder groups together, and has become established as the representative of airport pressure groups in Europe.

Industry response

The industry has responded to the growing interest in the environment. IATA has a long-standing task force, addressing the technical aspects of noise and emission stringency. More recently, a second task force with a broader responsibility, reflecting the wider general interest, has been formed. The Association of European Airlines has also established an environmental task force and other airlines are also placing the environment on the agenda (Orient Airline Association 1994). In a recent survey of members, IATA identified some 40 airlines with environmental focal points. The Airports Council International (ACI) also has set up environmental groups both on a global and European basis.

THE FUTURE

The twentieth century has seen truly amazing advances in flying technology. This has been reflected by major changes in the types of aircraft flown. However, there is a real question as to whether this pace of change will be maintained in the future. While engine and aircraft manufacturers will continue to introduce new technologies (see for example Martin 1991), it seems that these will appear mostly in overall shapes very similar to those of present-day aircraft. One aspect of the 'peace dividend' is that governments are reluctant to spend freely on the development of military aircraft, which have been a fruitful source of new technology for the civil aviation industry in the past.

The Boeing 777 will enter service within the next year. British Airways, one of the lead customers, took delivery of its first such aircraft in late 1995. It is anticipated that fuel efficiency will be measurably better than comparable existing aircraft. The British Airways aircraft will be powered by the new GE90 engine, incorporating a dual annular combustor which should lead to a significant decrease in NOx emissions. In early 1995, Swissair took delivery of its first Airbus 320 series aircraft with CFM56-BDAC engines, which incorporate very similar low NOx technology. However, it should be stated that there are some reservations about the relationship between certified NOx emissions, based on ground-level determination, and actual emissions of NOx at cruise altitudes.

The concept of an aircraft seating 600 or more and above (New Large Airliner, NLA) was instigated by British Airways, and is now under investigation by manufacturers such as Boeing, McDonnell Douglas, and Airbus. Such an aircraft, which is still at the conceptual stage, should make a contribution to relieving congestion at airports, for example, at Heathrow, where runway capacity is an important factor. At least one manufacturer has indicated confidence that noise levels can be achieved comparable to those of the Boeing 747–400.

Various studies have suggested that there may be a market for an approximately 300-seat supersonic airline (SST). Such an aircraft would fly subsonic over land and would have a range of some 6000–6500 miles in the stratosphere, flying around Mach 2.2. As with other new concepts such as the NLA, economic, technical, and environmental considerations are all important. The US National Aeronautics and Space Administration (see NASA 1992) is investigating the effect of possible designs of future SSTs on the stratosphere, and such an aircraft will not fly without meeting stringent environmental requirements. The current SST, Concorde, is not classified under Annex 16 of the Chicago Convention. Thus it is not considered for inclusion in Chapter 2 or Chapter 3 with respect to noise. The environmental impact is limited because of the small number of aircraft (13) in service.

It is worth mentioning one other concept—propfans, aircraft powered by engines with counter-rotating propellers. Such aircraft offer potential for improvements in fuel efficiency. Indeed, an aircraft has flown with such an engine. However, they are at present ruled out because of problems with noise and vibration both within the aircraft and transmitted to the ground.

These examples of new technology will only flourish if they are integrated with an environmentally effective infrastructure to support travel and tourism in the twentieth century.

SUMMARY AND CONCLUSIONS

The aviation industry has long been aware that it has an impact on the environment. As with many other industries, the nature of that impact has changed as our understanding of man's interactions with the environment has improved. Waste generation, congestion, and tourism are general environmental issues in which the activities of aviation play a part, alongside the more specific aspects of noise and emissions. To alleviate congestion, there is a clear need for improvements in infrastructure in the air and on the ground. Although the industry will continue to strive for technological improvements in noise, the capacity for managing at source is limited and this points more and more to stricter land-use policies.

We cannot be complacent in assuming that we have a thorough understanding of the impact of aviation. At the present time this applies particularly to the understanding of the impact of emissions at high altitudes, although new technology is being developed that will contribute to amelioration of such impacts. Significant research effort is being dedicated to improving our understanding of this area. At present, aviation makes a minor, but measurable and significant, contribution to carbon dioxide emissions. However, if predicted growth continues indefinitely and global warming and associated climatic effects remain a major environmental problem for the world, aviation will face challenges.

Although no thorough evaluation has been carried out, it would clearly be possible to use hydrogen, and possibly other fuels, for civil aviation. This begs the question of the environmental acceptability of the generation, supply, and use of hydrogen. Although a Tupolev 155 has flown with one engine fuelled by hydrogen, one key factor for aviation is the requirement for three times as much volume for the same energy content, a fact that would require major changes in aircraft design.

None of these possible advances is sufficient to turn aviation into a sustainable industry, at least according to some interpretations of the meaning of the term 'sustainable development'. It is easy to be cynical about an industry such as aviation or other forms of transport that are clearly having an impact on the environment. However, the challenge is to develop in a way that will match the long-term aspirations of society and protect the planet. Travel is high on the current list of aspirations and it is hard to see that pressure decreasing. Sustainable development is a concept for society as a whole and one for which the implications are not yet clear. For aviation there are long-term alternatives such as flying aircraft on hydrogen. The industry already has an outstanding record of achievement in response to environmental challenges and will continue to participate in the developing understanding of this important concept.

REFERENCES

Airfields Environmental Federation (1994). *Proceedings 'Aviation Growth and Environmental Sustainability'*. Airfields Environmental Federation, High Timber St, London.

Acher, L. (1993). *Aircraft emissions and the environment*. Oxford Institute for Energy Studies, Oxford.

Barrett, M. (1994). *Pollution control strategies for aircraft*. World Wide Fund for Nature, Godalming, Surrey.

British Airways (1991). British Airways environmental review—Heathrow and the world-wide flying operations. British Airways, Heathrow.

British Airways (1994). *Annual environment report*. British Airways, Heathrow.

Clewell, H. J., (1983). Ground contamination by fuel jettisoned from aircraft in flight. *Journal of Aircraft* **20**, 382–4.

CAA (Civil Aviation Authority) (1994). *Noise exposure contours for Heathrow Airport, 1991*. CAA, London.

Department of Environment (1994*a*). *Planning policy guidance: planning and noise*, PPG 24. HMSO, London.

Department of Environment (1994*b*). *Digest of environmental protection and water statistics*. HMSO, London.

Department of Transport (1992). *Report of a field study of aircraft noise and sleep disturbance*. Department of Transport, London.

Eber, S. (ed.) (1992). *Beyond the green horizon*. World Wide Fund for Nature, Godalming.

Egli, R. A. (1990). Nitrogen oxide emissions from air traffic. *Chimia* **44**, 369–71.

European Commission (1992). *Green paper on the impact of transport on the environment*. European Commission (DG VII), Brussels.

Heathrow Airport Limited (1994). *Environmental performance report*. Heathrow Airport Limited, Heathrow.

IATA (International Air Transport Association) (1991). *Air transport and the environment*. IATA, Geneva, Switzerland.

IATA (International Air Transport Association) (1993). *World air transport statistics*. IATA, Geneva, Switzerland.

ICCAIA (International Coordinating Council of Aerospace Industries Associations) (1994). *Aircraft noise design effect study—economic impact summary*. ICCAIA, Washington, DC.

IPCC (Intergovernmental Panel on Climate Change) (1994). *Radiative forcing of climate change*. IPCC and World Meteorological Organisation, United Nations Environment Programme, Geneva.

Japan Airlines (1994). *Environment friendly JAL*. Japan Airlines, Tokyo.

Johnson, C., Henshaw, J., and McInnes, G. (1992). Impact of aircraft and surface emissions of nitrogen oxides on tropospheric ozone and global warming. *Nature* **355**, 69–71.

Martin, R. L. (1991). *Air quality—an airplane manufacturer's view*, Air Quality Seminar. Birmingham Airport, Birmingham, UK.

Miles, D. E. (1994). The atmospheric impact of aircraft emissions, an issue for mutual EU activities between aeronautics and environment. In *Impact of emissions from aircraft and spacecraft upon the atmosphere*, proceedings of an international scientific colloquium (ed. U. Schumann and D. Wurzel), pp. 3–8. Deutsche Forschungsanstalt für Luft- und Raumfahrt, Cologne.

NASA (National Aeronautics and Space Administration) (1992). *The atmospheric effects of stratospheric aircraft—a first program report*. NASA, Office of Space Science and Applications, Washington, DC.

Orient Airlines Association (1994). Economics and the environment—the airline management challenge, Conference, Kuala Lumpur. OAA, Manila, The Philippines.

Raper, D. W. and Longhurst, J. W. S. (1990). *The impact of airport operations on air quality*, National Society for Clean Air Workshop, Newcastle, 1990. National Society for Clean Air, Brighton.

Royal Commission on Environmental Pollution (1994). *Transport and the environment*, 18th report of the Royal Commission on Environmental Pollution. HMSO, London.

Schumann, U. (1994). Impact of emissions from aircraft and spacecraft upon the atmosphere—an introduction. In *Impact of emissions from aircraft and spacecraft upon the atmosphere*, procedings of an international scientific colloquium (ed. U. Schumann and D. Wurzel), pp. 8–13. Deutsche Forschungsanstalt für Luft- und Raumfahrt, Cologne.

Schumann, U. and Wurzel, D. (ed.) (1994). *Impact of emissions from aircraft and spacecraft upon the atmosphere*, proceedings of an international scientific colloquium. Deutsche Forschungsanstalt für Luft- und Raumfahrt, Cologne.

Smith, M. J. T. (1989). *Aircraft noise*. Cambridge University Press, Cambridge.

Swissair (1992). *Environmental audit, 1992*. Swissair, Zurich.

Vedantham, A. and Oppenheimer M. (1994). *Aircraft emissions and the global atmosphere*. Environmental Defense Fund, New York.

WTTC (World Travel and Tourism Council) (1994). *Travel and tourism—the world's largest industry*. WTTC, Chausée de la Hulpe, Brussels.

8

The oil industry, transport, and the environment

Rodney Chase

Mr Rodney Chase is now Chief Executive Officer of British Petroleum (BP) Exploration. When he delivered this lecture, he was the Managing Director responsible for communications, external affairs, and the environment in British Petroleum plc, the company to which he has devoted his entire career.

Rodney Chase has exceptional breadth of experience in the oil industry, having held posts within both the Upstream (exploration and production) and the Downstream (refining and marketing) divisions of BP on three continents: Australasia, Europe, and North America. He has served the company in BP Shipping, in Refining and Marketing, in Distribution, in Oil Trading, and in Gas, as well as holding corporate positions in Finance and Strategic Planning. As Chief Executive Officer of BP Finance and Group Treasurer, Mr Chase played a key role in the company's acquisition of The Standard Oil Company and Britoil, as well as in the divestment of BP's mining interests.

Currently, in addition to his responsibilities for exploration, Rodney Chase has Board responsibility for the Western Hemisphere. He is Chairman of the World Business Council for Sustainable Development (WBCSD) and a Member of the UK Advisory Committee on Business and the Environment (ACBE); he also belongs to the UK Round Table on Sustainable Development.

INTRODUCTION

The title of this series of Linacre Lectures—which BP was delighted to sponsor—is certainly topical. In February 1995 the Secretary of State for Transport, Dr Brian Mawhinney, launched what he described as the great transport debate'. Given that our series of lectures was planned long before Dr Mawhinney's translation to the Cabinet, I think that we have more than demonstrated our prescience.

My specific topic is 'Transport and the environment as it concerns the oil industry'. One could hardly have chosen a more all-embracing title, but I hope it is worthy of Thomas Linacre. As the founder and first President of the Royal College of Physicians, Linacre would have approached environmental problems primarily in terms of their health implications—which, when one thinks about it, is not very different from our attitude today. And he, too, would have bracketed transport and the environment together. One of the big health problems of his time was tetanus, caused largely by the 'exhaust' emissions of the then means of transport—every bit as smelly and unhealthy as anything we are used to, although admittedly solid.

However, at least Thomas Linacre did not have to worry about the oil industry. That is my added complication, since to bring oil, transport, and the environment into a single chapter could require a vast canvas indeed. Perhaps then I should begin by stating what I am going to discuss—and what I am *not* going to discuss.

I aim to clarify the oil industry's contribution to solving the problem of transport emissions. We recognize this as an area of public concern and, as I hope to show, we are already deeply involved in the process of bringing about improvements. I shall bring to your attention the dangers of acting precipitately, without adequate scientific knowledge or clear objectives. I shall argue that it is possible to meet world health standards in a way that is compatible with scientific and economic realities. I shall suggest some approaches to transport policy that accord with the oil industry's general approach to energy and environmental policies as a whole. In this context, I hope to show that a responsible approach to both transport and the environment is entirely consistent with our commercial obligations to shareholders and customers.

What I will not do, however, is claim that the oil industry is equipped to decide priorities in what are proper areas of concern for *public* policy. Take, for example, health. The oil industry is eager and willing to deepen knowledge in this area. Indeed, through the Institute of Petroleum, the oil industry is currently funding a study that will greatly add to the understanding of the relationship between exposure to very low levels of benzene and the incidence of leukaemia. We await its findings. But, in the final analysis, it is the role of government, on behalf of society, to decide what does, or does not, constitute a health hazard. We shall do everything necessary, and often more than is necessary, to meet health standards and regulations. But, given that no activity is risk-free, society must decide where to draw the line between minimizing the risk, and seeking to eliminate it altogether.

Equally, it is not for the oil industry to adjudge the public debate between private or public transport, whether under privatized or state ownership. We may well have our own views as to what constitutes an effective transport strategy and, as an industry with obvious interests in transport, we shall be ready to make these views known. But we do not claim any particular expertise in this area. Our commercial behaviour *reacts* to transport policy decisions; it does not seek to dictate them.

I must also make clear at the outset what I understand by the word 'environment', since it often means different things to different people. The oil industry tends to see the environment primarily in terms of its interaction with, and impact on, air, land, and water. We are anxious that our operations should inflict as little damage as possible on the use and enjoyment of physical amenities such as open space, countryside, and natural surroundings. We have an overriding objective—to avoid environmental accidents and disasters. But, apart from this, our main practical concern is with the local, regional, and global atmospheric consequences of producing and consuming our products.

Transport is, of course, a vital component of this whole question, but the priorities here are probably different. Even if a fuel were to be invented tomorrow that was capable, at minimum cost, of propelling vehicles without any detrimental atmospheric effects whatsoever, there would still remain a myriad of transport issues to be tackled such as congestion and the impact of roads and motorways on local populations. Such problems arise, irrespective of the substances that may, or may not, be emitted into the atmosphere. It is important to remember, therefore, that in many respects the oil industry is a minor player in a much larger debate.

Indeed, to some extent, emissions are a symptom, rather than a cause, of the problem. Reduce congestion and you help to reduce air pollution. Encourage public transport, and you cut carbon dioxide (CO_2) emissions. Reducing air pollution *per se*, however, does nothing to reduce congestion—unless you take the view that we shall all take to our bicycles once urban air is pleasant to breathe. You might also argue, however, that 'civilizing the car' could lead to an increase in its use. Cleaner and quieter cars might lead to more people wanting to drive them, but then, of course, the motorist would exacerbate those very congestion pressures, which, irrespective of pollution, would conspire to make such growth unsustainable. The point I am making is that transport policy—and, particularly, transport policy aimed at reducing congestion—is distinct from pollution questions generally, even though there is a very close connection.

I notice that most of the contributors to this volume could easily be described as 'interested' witnesses. In that regard, I suppose I am no exception, so I should explain how I approach this subject. I have no doubt that most people assume the oil industry to have a vested interest in as many people as possible driving cars, and therefore consuming our products.

Would it not be natural, they ask, for oil companies to feel threatened by the development of alternative fuels or by technological breakthroughs that make the combustion engine obsolete? The honest answer to this depends upon the time-scale you are considering. Of course, our current business is largely about producing and selling oil. If some wonder product were to come on to the market overnight to make oil as useful (or as useless) as charcoal, it would present us with problems—to say the least! But in realistic terms, I find it very difficult to see anything happening in technological terms to displace oil as a transport fuel, in a way that could transform the environmental agenda, for the foreseeable future.

Even if a technological substitute for oil existed, the existing car fleet could not be replaced overnight. The current European Union (EU) view is that it would take between 10 and 15 years for this to happen. And it is a general rule of thumb that few of the alternatives would produce improvements so quickly or effectively as repairing, and properly maintaining, all those old and badly maintained vehicles currently on the road. Given that some 80 per cent of today's emissions are accounted for by 10–20 per cent of vehicles, this really should be the first area to tackle as an urgent necessity.

Of course, I am not denying that technological revolutions may well occur during the next century that will have a major impact upon the oil industry. But they are too far off to justify a complacent, or starry-eyed, approach today. It is difficult to envisage anything happening over the next 20 or 30 years to make traditional fuels obsolete. This need not be a source of concern. We have not yet reached the limit of improvements in the petrol combustion engine. Great strides are still possible—and I would submit that this is an objective we should concentrate upon.

But, before I do so, let me refer briefly to energy efficiency and how more responsible motoring could help redress the problems we face.

ENERGY EFFICIENCY

The oil industry accepts and welcomes its role in helping our customers to consume our products efficiently and without waste. For example, we are always ready to advise our customers on how to reduce their fuel consumption. BP is particularly interested in the customers of the future. We have a programme entitled 'Challenge to Youth' that aims, in part, to foster better road and traffic awareness amongst young people. The scheme embraces an under-17 car club, road-safety instructions in schools, and a 'build-a-car' competition, often with fuel-efficiency as a main criterion of success.

This is in keeping with our approach in other fields. For instance, BP has for many years offered advice on home insulation, and on how to promote energy efficiency generally in both residential and commercial buildings. We have our own subsidiary company—BP Energy—which specializes in the management of industrial heat and power plant for improved energy efficiency, and in the design and construction of combined heat and power plant. We estimate that BP Energy has enabled its customers to improve the energy efficiency of existing plant by between 5 and 15 per cent through better management techniques alone.

I trust that these examples illustrate our commitment to energy efficiency. But we should not be under any illusions. Even if achievements in this exceed all expectations, they would never be enough to make the environmental problems associated with transport disappear.

RISING DEMAND FOR TRANSPORT

One reason for this is that transport demand is growing. While it is true that many of the markets in which the oil industry operates are mature and highly competitive, demand elsewhere is forecast to continue to increase significantly. Forecasts vary, but some suggest that the world's car population will have risen from what we believe to be some 500 million today to 550 million (and possibly even more) by the year 2000. Not that we should accept such forecasts too readily. One of the pitfalls surrounding the 'transport and environment' debate is the number of statistical black holes that exist. In fact, there are no precise figures as to how many cars there are on the world's roads; the figure above represents a best shot, but is by no means certain. And the uncertainties are even greater when assessing the distance cars travel per year on average, or how many miles per gallon they achieve. This, incidentally, is one of the problems of adopting a rational approach based on good science—often, we lack sufficient data to do so. Nevertheless, I think we can safely conclude that road travel is on a rising trend. The growth potential in road fuel consumption in both India and China is vast. Coupled with South-east Asia, it underlines how attention will need to concentrate on these countries as much, if not more, than on Europe and the United States if environmental issues are to be tackled effectively.

Not that we can forget what's happening in our own backyard. In the UK alone, passenger travel by all modes has increased by over 40 per cent since 1979. Car travel in the UK has more than tripled. There has been a tenfold increase in distances travelled by car over the last 40 years. Nearly 70 per cent of UK households now have the regular use of a car, although it's worth remembering that car ownership in the United Kingdom is still lower than in France, Sweden, or Germany—let alone the USA. Taking the EU as a whole, it has been estimated that the number of cars will grow by 25–30 per cent during this current decade, based on trends since 1975.

In the UK, much attention is given to the split between road and rail travel, and the desire for private, as opposed to public, transport. This debate is placed in its proper perspective by data released by the Department of Transport that states that even a 50 per cent increase in rail travel would still only be equivalent to as little as a 3 per cent reduction in travel by road. That is less than the average annual growth in road transport over the last 10 years. A similar point is made for freight. According to the Department, a 50 per cent increase in rail freight would be equivalent to only a 5 per cent reduction in road freight. And, while attention normally centres upon the private motorist's reluctance to give up his car, it is important to remember that freight transport—and particularly road freight—plays a pivotal role in making a reality of Europe's 'single market'.

Dr Mawhinney's speech in February 1995 developed this point. He reminded us that each year about 2000 million tonnes of goods are moved in this country alone. Over 160 million tonnes are exported, and 99 per cent of our exports by weight leave the country by sea. But often these goods travel long distances to get to port, mostly by road. Up to now, we have witnessed a long-term decline in rail's share of the freight market. But Dr Mawhinney pointed to the significant quantities of rail freight that are beginning to go through the Channel Tunnel, and that could possibly take some 400 000 lorries a year off the roads within 2 to 3 years. So, perhaps rail has a real chance of capturing some of the market it has lost. What is certain is that a 'Europe without frontiers' will mean very little if people can't move goods across them.

But, if transport policy for freight is crucial for the economy, it is personal mobility that concerns most voters; the private car may attract most expert criticism, but it is the form of transport that most people are loath to give up. And not just in the UK. The popularity of the car is clearly evident in other developed countries as well. And while it is untrue to say that there is consumer aversion to public transport—it is of interest, for example, that UK bus mileage outside London has increased by over 20 per cent since 1986 and that over 80 per cent of commuters to central London rely on public transport—it is very difficult to believe that a significant move away from cars can be achieved without draconian measures. I merely remark on these statistics to underline the economic significance of transport and the importance that most people in the developed world—and a growing number in the newly developing world—attach to personal mobility. That is why transport policy so often impinges upon fundamental issues of personal freedom. It is not the job of an oil company to do anything to curtail this freedom. That is a nettle for the politicians to grasp, if they choose to do so. Our role is to supply the fuel required by our customers—and to supply it as efficiently and responsibly as possible. Clearly, personal transport yields enormous benefits to us as individuals, particularly if we live in rural areas. Even the most enlightened and efficient mass transport system imaginable would be hard put to retain the flexibility and convenience afforded by the private car. The oil industry does not deny the long-term strategic need for a sustainable transport policy. Along with so many other industries, our business requires an efficient transport network, and we are as much threatened by 'grid-lock' as any other commercial concern. But, given the right framework in which consumer preferences and costs are registered, our job is to ensure that oil is available for transport in the quantities and *qualities* that are demanded.

There are two major issues that arise in this context. One is the efficiency of vehicles, and the extent to which changes in the design and manufacture of cars can have a beneficial impact upon energy conservation and the environment. The other is the quality of the fuels themselves—whether they can be improved so that they are cleaner and more efficient. And in both cases, there are cost implications—for the customer, and for society generally. The design of vehicles is primarily an issue for the motor industry, while the chemical composition of fuels falls largely into our camp. But, of course, this is an academic distinction, and both industries must act in concert if an optimum solution is to be found. We need a 'systems approach'—by which I mean we need to examine fuels and vehicles as part of a single exercise. The oil industry is, and wants to remain, a part of this process.

EMISSIONS

Let me turn to emissions first. The emphasis needs to be upon 'good science' and 'practical results'. Much has already been achieved, as I will show. There are a range of possible measures yet to be taken. But they must be subjected to rigorous cost-effective analysis, since we are very close to the point where further improvements in this specific area can only be introduced at great cost and for little effect.

The Department of the Environment (1995) has forecast dramatic reductions in air emissions over the next 15 years, as a consequence of a wide range of measures already taken. The Royal Commission on Environmental Pollution (1994) has recognized that emission levels are falling, although—to be fair—they want further improvements. In BP, we actually believe that the Department's forecast of a 36 per cent reduction in total volatile organic compounds (VOCs) emissions will be exceeded. This is primarily because of efficiency improvements in refineries that exceed the Department of Environment's projections and because the chemical industry's performance in this area has surpassed expectations. But the projected reduction of transport emissions also now looks conservative, largely thanks to the increasing fuel efficiency of vehicles. By the end of the century, some 60 per cent of UK vehicles will have been fitted with both catalytic converters and small carbon canisters. This is crucial to reducing vehicle emissions, because this technology cuts exhaust emissions, including VOC emissions during normal running, by up to 90 per cent. Taking action now to reduce emissions over and above those already projected could prove very expensive if done in the wrong way. The UK oil industry alone has already spent some £350 million in capital expenditure on the Stage 1 vapour recovery programme. This is an EU initiative to control emissions during tanker loading at terminals and during their discharge at retail sites. The programme also involves an annual operating cost of some £25 million.

To go further is technically possible, provided the industry is given time to make the necessary investments. But that still begs the question whether a small improvement in urban air quality could not be achieved in better ways. Are we at least not obliged to ensure there are no other less costly and equally effective solutions available? This is precisely what the European oil and auto industries are studying at present under the auspices of the EU initiative on Air Quality, Emissions, Fuels and Engine Technologies, which is led by the Commission. The aim of the initiative is to come up with a proposal later this year on how to meet air quality standards in the year 2000.

As I have intimated, much progress has already been made in Europe. For example, if we look at the EU as a whole, it is now possible to say that:

- unleaded gasoline and low sulfur diesel are increasingly available throughout Europe;
- all new gasoline-powered motor vehicles are equipped with three way catalytic converters and electronic ignition systems, and require unleaded fuel;
- a new passenger car bought in 1993 and subsequent years will emit no lead, approximately 93 per cent less carbon monoxide, and 85 per cent less hydrocarbons and nitrogen oxides than a new car bought in 1970;

- further progress is being made on diesel engine improvements;
- industry research based on the car fleet size and composition, and on the number of kilometres driven on average per year, predicts that between 1992 and 2010—and as a result of steps already taken—the amount of pollutants in tons per year due to passenger car traffic will decrease by approximately 75 per cent.

One of the features of those developments is that they involve the oil and motor industries working in good faith to come up with solutions that are technically feasible and that offer best value in terms of economic costs. The European Commission has the right of proposal; but the involvement and commitment of the European and national parliaments is also vital, because it is often necessary to enshrine in legislation what is expected of business. In any industry—and the oil industry is no exception—measures of an environmental nature arise that are worthy of implementation, but that, for reasons of cost, no company can adopt unilaterally.

This is not always the case. Sometimes, companies take an environmental initiative in order to steal a march upon their competitors, though often this is short-lived. But there is an equal danger that a single initiative by a company will prove environmentally ineffective, and dangerous commercially. If such an initiative is nevertheless desirable, members of the industry will often be to the fore in requesting government to lay down a minimum standard or regulation to ensure that everyone plays to the same rules.

This also applies very much to shipping, where similar issues to those I have been discussing arise. In the last decade or so, attention has focused on shipping's role as a contributor to air pollution. Many feel that the time has come for international shipping to put its own house in order. Because of the international nature of shipping, truly effective controls—which do not distort competitive positions—can only be achieved through the activity of an authoritative and international body. It is for this reason that air pollution from shipping has been under active consideration at the International Maritime Organization (IMO) since 1988.

Halons, chlorofluorocarbons (CFCs), VOCs, and nitrogen oxides (NOx) have all been under the spotlight. But the issue that has the greatest impact on the oil industry, and the one that has generated greatest controversy at the IMO, is that of sulfur dioxide (SO_2) emissions. As so often with transport issues, the problem is in differentiating between local and global concerns. Clearly, the environmental impacts of SO_2 emissions in mid-ocean are very different from those in ports, or from other land-based sources.

The oil companies' European organization for environment, health, and safety, known as CONCAWE (Conservation of Clean Air and Water in

Europe), undertook a series of comprehensive studies to identify the environmental impact of SO_2 emissions from shipping, and to evaluate the cost of producing low-sulfur bunker fuels. The conclusion was that the cost could be as high as $20 billion, and that the environmental benefits of a global reduction would be very small. On the other hand, the CONCAWE studies demonstrated that, in certain areas, ship-sourced emissions of SO_2 can exacerbate identified environmental problems significantly—which is why the oil industry supports what has become known as the IMO's 'Special Areas' approach. The intention is that restrictions on the sulfur content of bunker fuels should be imposed in areas where there is a demonstrated need on environmental grounds for such restrictions. Once again, this is a complex area. As with road transport emissions, it is easy to spend a vast amount of money for little, or no, effect. There will always be those who argue that 'more could be done'. But the oil industry accepts its obligations in this area and is working with the IMO to ensure that whatever is finally agreed is practical, effective, and applied internationally.

Uniform enforcement is always critical. It is, incidentally, a partial answer to the question 'Isn't what you've done to improve air emissions the consequence of government intervention and legislation, without which you'd have been happy enough to do nothing?' The answer to the first part of the question is a qualified 'yes'. Sometimes we do wait for government intentions to be clarified before taking action. However, the accusation that we would have been happy to do nothing indefinitely is unfounded. It is not in the oil industry's interests that the use of oil should carry with it negative connotations. Buying petrol may not be the most exciting activity in the world, but we have no interest in encouraging people to feel guilty about purchasing our products.

However, occasions exist when a single company can do little to achieve major improvements on its own initiative, and statutory standards, assuming full compliance, do at least ensure that the costs are shared throughout the industry. When these are coupled with implications for competitiveness, it is not difficult to see why the oil industry often calls for general government guidelines, provided these are neither prescriptive nor irrational.

Before I leave the issue of air emissions, I want to make two further points. The first concerns our own operations. I have been focusing on the motor car, but there is an extent to which oil industry activities themselves contribute to air emissions. For example, the oil industry distributes many of its own products by road. We have a commercial interest in tackling congestion and, on grounds of cost, we wish to keep the time our vehicles spend on the road to a minimum. We've already made progress.

For example, BP tankers now travel fewer miles than 4 years ago in spite of having fewer terminals from which to distribute petrol and diesel.

My second point concerns reformulated gasoline, which is often held up as an answer to the problem of air emissions, and which has strong champions in the USA—or at least used to before US motorists began to flex their political muscles. The costs of fuel reformulation in the US are very high, and it is far from certain whether they are justified by results. Indeed, the US approach of allowing regions to opt in and out of reformulated gasoline provides a classic example of industrial policy at its worst. US oil companies, having invested hundreds of millions of dollars in order to comply with the legislation on reformulated gasoline, now find themselves providing an expensive product of debatable environmental value for which there is precious little demand.

There are growing signs of a consumer and political revolt in many parts of the United States. Reformulated gasoline prices are some 5 to 14 cents per gallon higher than those for conventional petrol. While no more than the oil industry predicted, this has not made the price any more palatable to consumers, many of whom question the rationale behind its introduction. Many counties, in states like New York and Pennsylvania, have already opted to pull out of voluntary participation in the programme. Not only has this caused chaos for the industry, but it also demonstrates how easily resources can be wasted for no environmental benefit. I can think of no better example than this to illustrate my fundamental point. Industry will always do what is required of it by legislation. But if that legislation is ill-conceived, the risk of a scandalous waste of resources is very high. And those resources might have been deployed elsewhere for *real* environmental benefit.

The reaction to reformulated gasoline may be part of a much wider trend. I quote from an article from the *Wall Street Journal*—not because I agree necessarily with the sentiments, but because I think it illustrates an interesting development:

> The Clean Air Act . . . is under attack across the country . . . Drivers in the New York area can listen to radio talk-show hosts gripe about how reformulated gasoline—the more expensive but cleaner-burning fuel now sold in the nation's smoggiest cities—may be harming their engines and is generally a dumb idea dreamed up by bureaucrats . . . Cars have emerged right up there with guns as a Constitutional right.

Personally, I think the 'right' to drive is far more defensible than the 'right' to shoot! But for our purposes what is important is that such things are being written. Many people are predicting a counterrevolution in the

USA's environmental approach, caused in part by consumer resistance to higher prices and political dislike of overregulation. It will be fascinating to see whether this takes place and, if so, what implications it has for Europe.

My final point on air emissions concerns alternative fuels. The problem here is that all the main contenders have their strengths and weaknesses. No single fuel solves all the problems. And a great deal depends upon the time-scale involved. For example, there is no doubt that electric vehicles are very quiet and are emission-free at the tail-pipe. If their current technological limitations could be overcome, many would find them the perfect answer to inner city pollution. The fact is, however, that the hoped-for breakthrough in battery technology has not yet happened, despite large research and development expenditure. Equally, it is worth remembering that the environmental consequences of electric cars are not totally benign, especially when considered from the cradle-to-grave perspective. It all depends upon the type of fuel used to generate the power which charges the batteries. Only nuclear-generated electricity would provide electric vehicles with an overwhelming environmental advantage. And, as we all know, that is an option that would alarm many people.

Gas-fuelled vehicles, on the other hand, are now established technology. They are not cheap compared to petrol or diesel, and have limited range. Heavy, bulky cylinders are required to carry the gas. But, from an environmental point of view, they are beneficial. I have no doubt that gas-fuelled fleet vehicles, particularly small delivery vans and taxis, will gain in popularity in areas where gas is readily available.

But there is no reason at present to expect either electric or gas-fuelled vehicles to displace the combustion engine fuelled by petrol or diesel, because other technical possibilities—such as the production of ultra-low emission vehicles (ULEVs for short)—will allow conventionally fuelled cars to be driven for much of the next century with even lower emissions than those of the most modern cars at present. For example, Honda announced at this year's Los Angeles Motor Show that it was the first manufacturer to have a gasoline engine certified as meeting California's ultra-low emissions standard due in the year 2000. Based on its existing 2.2-litre Accord engine, and fuelled by Californian reformulated gasoline, it uses variable valve-timing, fast-warm catalyst, and precise computerized management of the fuel-to-air ratio. It is claimed that emissions from this vehicle are cut by an average of 90 per cent from 1994 levels.

My instinct is that we should follow this approach and conclude that perhaps the best solution is to work towards the optimization of diesel or petrol and the engine *as an integrated system*. Both need to be considered simultaneously within an integrated approach to the problem. However, this will not overcome the transport problem. In the final part of this chapter, I would like to give an oil company response to some of these issues. In particular, since they are our customers, are we happy to see the private motorist driven off the road? An even more pertinent question is whether the private motorist is happy to be driven off the road? That, in the end, is what will determine the answer.

TRANSPORT POLICY

Our approach to transport policy is similar to our approach in other areas. We believe it is important to understand as fully as possible the true costs of each mode of transport, and to ensure that these are known to, and borne by, the customer. This is a question beloved of economists. It is also a question which goes to the heart of the environmental debate. We must continue to strive for agreement on how the relative benefits and costs of different transport modes can be identified and passed on to users. It is also important that realistic choices should be available. There is a danger in penalizing cars if no economic alternative exists.

In this context, one question facing policy-makers is whether economic instruments are able to guide consumers sufficiently quickly to the most favourable transport mix, or whether government needs to use its powers to achieve a more rapid and planned outcome. It is of significance that the European auto-oil programme has, as part of its work, an evaluation programme of various non-technical measures to influence transport decisions with a view to reducing air emissions.

Should vehicle ownership and/or vehicle use be made more expensive? What part can improved traffic management play to reduce congestion by, for example, increasing parking charges or banning vehicles from inner city areas? To what extent should taxpayers' money be used to make public transport more attractive? The European Commission is currently reviewing all these questions, and industry is fully engaged in the consultative process.

So far as BP is concerned, we have expressed our support for a number of practical measures that fall under the aegis of transport policy. The first is the effective enforcement of maintenance standards by improved vehicle inspection. A mixture of penalties and more effective surveillance is called for. BP also supports improved traffic management and information systems. Congestion is bad for air emissions, and parking and traffic regulations within cities have a major part to play. So does the siting of supermarkets and offices.

Another reason for congestion is inadequate information, and we believe that this is an important area to improve. If every road-user had an effective early-warning system of traffic ahead, some passengers might either re-route their journeys or postpone them. I have a lot of sympathy for what Simon Jenkins wrote in *The Spectator*: 'We take our car onto a stretch of road as long as, but only as long as, it gets us to a destination by a certain time . . . Traffic achieves equilibrium at a certain level of congestion . . . What destroys this equilibrium is an accident or unexpected incident . . . ' Even if Simon Jenkins is only partially right, I suspect we should do all we can to encourage as a matter of routine the introduction of technology to enable drivers of cars to have reliable and instant travel information during the course of their journeys.

BP also supports the encouragement of responsible driving—I have already mentioned our work with young people, in this connection.

Public transport has 'a role to play', although its importance may be overstated. If scientific opinion were to agree upon the nature of the threat posed by global warming, and were to conclude that this was every bit as dangerous as the most pessimistic forecasts predict, then governments might feel compelled to force people on to public transport for certain kinds of journeys—and that, in practice, would probably mean buses. They could achieve this in any number of ways—by taxing cars out of existence; by making it illegal to own more than one car; or by restricting the number of vehicles that any household could park on a public road. But, bearing in mind the growth expected in India and China, these self-imposed sacrifices, while reducing traffic queues, would have a minimal effect on global CO_2 emissions.

A 'no-regrets' approach to global warming would draw back from radical action to outlaw private motoring unless and until it became inescapable, and would instead favour an approach to transport policy whereby the car bears its full share of its costs, and the passenger is left to decide whether or not to pay them. This is easier said than done, because of the difficulty in defining such costs. A growing body of opinion, however, argues that road pricing could help bring more into balance the supply/demand pressures for road space, particularly in congested areas. Equally, if only there were better integration of roads, parking, and public transport, more people might be prepared to rely on trains and buses for the bulk of their journey. For many, parking restrictions around stations are an obstacle to their use—their removal could encourage more people to take the train and only use their car for the short journey to and from the station.

But we all have our pet schemes. It is important to adopt a rational and integrated approach. And here, our policy in other areas may be of relevance.

For example, when debating energy policy, the oil industry normally opposes initiatives that seek to plan arbitrarily the energy 'mix' in advance. Instinctively, I suspect that most businessmen are equally suspicious of attempts to decide in advance the breakdown between various modes of transport, a question that hinges on whether or not transport markets can be made to function effectively. And that is a matter upon which the oil industry is not especially qualified to make a judgement. In reality, I am sure that there is a long way to go before an 'integrated transport policy' becomes detached from political scrutiny and controversy.

CONCLUSION

In this chapter I have attempted to approach some of the questions that critics of the oil industry might be expected to raise. Our first responsibility is to provide the energy which society demands. But, if and when governments choose to influence the shape of this demand, our wish is to be part of the debate in order to help arrive at a rational solution. We are committed to a positive and participatory role. Our main responsibility, along with our colleagues in the car industry, is to help identify the options and alternatives—and to beg for consistency and time to plan, once a decision has been taken.

However, we are well used to responding to new demands and to new social standards. Being 'in business' means in part anticipating, and satisfying, new 'likes' and 'dislikes'. Often, this calls for decisions well in advance of the event. These can be very costly, if misjudged, but competitive advantage comes to those who judge the mood aright. If society chooses to minimize the environmental impact of travel—either through higher environmental standards or through direct market intervention—then we regard this more as an opportunity than as a threat. Responding to new social pressures is a fact of business life, and not restricted to transport. Our biggest concern is that an emotional and unscientific approach in this, or any other area, will militate against the search for cost-effective and practical solutions. That *would* hurt the oil industry. But we wouldn't be alone. It would hurt the social and economic life of Europe as well, and harm Europe's competitive position in the world's markets. We ought to be able to see our way through many of these problems. We certainly have a duty and a common interest to do so.

REFERENCES

Department of the Environment (1995). *Air quality*. HMSO, London.
Royal Commission on Environmental Pollution (1994). *Transport and the environment*, 18th report of the Royal Commission on Environmental Pollution. HMSO, London.

Index